Oxford International Primary Science

Terry Hudson

Alan Haigh
Debbie Roberts
Geraldine Shaw

Language consultants:
John McMahon
Liz McMahon

3

OXFORD
UNIVERSITY PRESS

OXFORD
UNIVERSITY PRESS

Great Clarendon Street, Oxford, OX2 6DP, United Kingdom

Oxford University Press is a department of the University of Oxford. It furthers the University's objective of excellence in research, scholarship, and education by publishing worldwide. Oxford is a registered trade mark of Oxford University Press in the UK and in certain other countries

(c) Terry Hudson, Alan Haigh, Debbie Roberts, Geraldine Shaw 2014

The moral rights of the authors have been asserted

First published in 2014

All rights reserved. No part of this publication may be reproduced, stored in a retrieval system, or transmitted, in any form or by any means, without the prior permission in writing of Oxford University Press, or as expressly permitted by law, by licence or under terms agreed with the appropriate reprographics rights organization. Enquiries concerning reproduction outside the scope of the above should be sent to the Rights Department, Oxford University Press, at the address above.

You must not circulate this work in any other form and you must impose this same condition on any acquirer

British Library Cataloguing in Publication Data
Data available

978-0-19-839479-2

10 9 8 7 6 5 4 3

MIX
Paper from responsible sources
FSC® C007785

Paper used in the production of this book is a natural, recyclable product made from wood grown in sustainable forests. The manufacturing process conforms to the environmental regulations of the country of origin.

Printed in The United Kingdom

The questions, example answers, marks awarded and comments that appear in this book were written by the author(s).
In examination, the way marks would be awarded to answers like these may be different.

Acknowledgements
The publishers would like to thank the following for permissions to use their photographs:

Cover photo: Frans Lanting/Corbis, P04_05: David Jenkins/Robert Harding World Imagery/Corbis/Image Library, P5a: Paul Maguire/Shutterstock, P5b: Heiko Kiera/Shutterstock, P5c: James Laurie/Shutterstock, P5d: CreativeNature.nl/Shutterstock, P6a: Shutterstock, P6b: Viorel Sima/Shutterstock, P6c: Shutterstock, P6d: Jo Crebbin/Shutterstock, P6e: Shutterstock, P7a: Gubin Yury/Shutterstock, P7b: Shutterstock, P9a: Gerard LACZ/ NHPA/Photoshot, P9c: Shutterstock, P10: Ferderic B/Shutterstock, P11a: Anton Ivanov/Shutterstock, P11b: Shutterstock, P12a: Corey Nolen/Getty Images, P12b: Beverly Joubert/National Geographic/Getty Images, P12c: Michal Ninger/Shutterstock, P13a: Shutterstock, P13b: Dmitry Kalinovsky/Shutterstock, P13c: Alexey Kuznetsov/123RF, P13d: Masterfile, P14: Beverly Joubert/National Geographic/Getty Images, P15a: Pashin Georgiy/Shutterstock, P15b: Shutterstock, P16a: iStock.com, P16b: Shutterstock, P16c: Shutterstock, P16d: Ron Levine/Digital Vision/Getty Images, P16e: Shutterstock, P17: Shutterstock, P18a: Don Johnston/All Canada Photos/Corbis/Image Library, P18b: Ingo Arndt/Minden Pictures/Corbis/Image Library, P21a: Kitch Bain/Shutterstock, P21b: Shutterstock, P21c: Shutterstock, P21d: Dirk Ercken/Shutterstock, P21e: Shutterstock, P22a: Masterfile, P22b: Jan M Downer/Science Photo Library, P22c: Aleksey Stemmer/Shutterstock, P22d: Jill Lang/Shutterstock, P24: Ivonne Wierink/Shutterstock, P26: Shutterstock, P27a: Heiko Kiera/Shutterstock, P27b: Jo Crebbin/Shutterstock, P28_29: Masterfile, P29a: Shutterstock, P29b: Dreamstime.com, P31a: Evgeny Karandaev/Shutterstock, P31b: Shutterstock, P31c: Ingram/OUP, P31d: Todd Taulman/Shutterstock, P31e: Shutterstock, P31f: Pavel Kudryavtsev/Dreamstime.com, P32a: Joseph Gough/Dreamstime.com, P32b: Shutterstock, P32c: F. Moscheni/Sheltered Images/Glow Images, P33: REX/KeystoneUSA-ZUMA, P36a: Shutterstock, P36b: Joel Arem/Science Photo Library, P38a: Philip and Karen Smith/Iconica/Getty Images, P38b: Kai Pfaffenbach/Corbis, P38c: Xiaoyang Liu/Corbis/Image Library, P39a: Dave King/Dorling Kindersley/Getty Images, P39b: Stoyanov Alexey/Shutterstock, P39c: Chaiyaphong Kitphaephaisan/123RF, P40a: Shutterstock, P40b: Gabrielle & Michel Therin-Weise/Robert Harding World/Image Library, P40c: Shutterstock, P41: Peter Stuckings/Lonely Planet Images/Getty Images, P42: Shutterstock, P44a: Shutterstock, P44b: Tischenko Irina/Shutterstock, P44c: Shutterstock, P44d: Shutterstock, P44e: Shutterstock, P45a: Shutterstock, P45b: Shutterstock, P46a: Shutterstock, P46b: Linda Hartley, P50_51: judywhite / Garden Photos.com, P52: Lili Graphie/Shutterstock, P54a: Shutterstock, P54b: Shutterstock, P56: Dudarev Mikhail/Shutterstock, P57a: Dr Jermy Burgess/Science Photo Library, P57b: Dr Keith Wheeler/Science Photo Library, P62: Martin Leigh/Oxford Scientific/Getty Images, P63: Sandra Ivany/The Image Bank/Getty Images, P64a: Quincy Russel, Mona Lisa Production/Science Photo Library, P64b: Sandra Ivany/The Image Bank/Getty Images, P66a: Shutterstock, P66b: Lukasz Janyst/Shutterstock, P68a: Shutterstock, P68b: Janno Loide/Shutterstock, P69: Tracing Tea/Shutterstock, P72_73: Dan Breckwoldt/Shutterstock, P73a: Li Wa/Shutterstock, P73b: PM Images/Iconica/Getty Images, P74a: Barrett&MacKay Barrett&MacKay/All Canada Photos/Getty Images, P74b: Charles Sichel-outcalt/Dreamstime.com, P75: Tom Wang/Shutterstock, P76: Andrew Lambert Photography/Science Photo Library, P77: NASA/Science Photo Library, P78a: Shutterstock, P78b: Liliya Kulianionak/Shutterstock, P78c: Shutterstock, P79a: Photo by Jurvetson (flickr), P79b: Catherine Murray/Shutterstock, P80a: Bruce Dale/National Geographic/Getty Images, P80b: Shutterstock, P82: Shutterstock, P84a: Bruno Rosa/Shutterstock, P84b: Shutterstock, P85a: Shutterstock, P85b: Dreamstime.com, P86: Suphatthra China/123RF, P87: Shutterstock, P88a: NASA/Science Photo Library, P88b: Shutterstock, P88c: Shutterstock, P90a: Jacek Chabraszewski/Shutterstock, P90b: Mauricio Avramow/Shutterstock, P90c: Brian Mitchell/Corbis/Image Library, P91a: Monkey Business Images/Shutterstock, P91b: mother image/cultura/Corbis/Image Library, P91c: Edward Lara/Shutterstock, P91d: Alena Ozerova/Shutterstock, P91e: Arvind Balaraman/Shutterstock, P91f: Shutterstock, P91g: Shutterstock, P91h: KidStock/Blend Images/Corbis/Image Library, P95: Mauro Fermariello/Science Photo Library, P98: Deyan Georgiev/Shutterstock, P100a: Lesya Dolyuk/Shutterstock, P100b: iStock.com, P100c: Shutterstock, P100d: Shutterstock, P102a: Jose Luis Pelaez, Inc./Corbis/Image Library, P102b: Shutterstock, P102c: Zaharescu Mihaela Catalina/Dreamstime.com, P102d: Artur Bogacki/Shutterstock, P102f: Imagemore Co., Ltd./Corbis/Image Library, P103: Alix Minde/Corbis, P104a: Shutterstock, P104b: Ilya Andriyanov/Shutterstock, P106a: Hero Images/Corbis/Image Library, P106b: Eastern Light Photography/Shutterstock, P108: Shutterstock, P109: Photofusion/UIG via Getty Images, P110a: Oliver Benn/The Image Bank/Getty Images, P110b: Shutterstock, P112a: Duncan Noakes/123RF, P112b: Marcin Pawinski/Shutterstock, P113: Shutterstock, P114: Shutterstock, P116_117: John Lund/Tiffany Schoepp/Blend Images/Corbis/Image Library, P117: Forster Forest/Shutterstock, P118a: R S Vivek/Dreamstime.com, P118b: Skydive Erick/Shutterstock, P118c: Shutterstock, P118d: Shutterstock, P118e: Levent Konuk/Shutterstock, P119: Paul J. Richards/AFP/Getty Images, P122: Biophoto Associates/Science Photo Lbirary, P124a: Shutterstock, P124b: Shutterstock, P125: Steve Klaver/Star Ledger/Corbis/Image Library, P126: Purestock/SuperStock/Corbis/Image Library, P128: Masterfile, P130a: Ivonne Wierink/Shutterstock, P130b: Monica Schroeder / Science Source, P132: Ermolaev Alexander/Shutterstock, P135: Purestock/SuperStock/Corbis/Image Library, P137: Shutterstock.

Although we have made every effort to trace and contact all copyright holders before publication this has not been possible in all cases. If notified, the publisher will rectify any errors or omissions at the earliest opportunity.

Links to third party websites are provided by Oxford in good faith and for information only. Oxford disclaims any responsibility for the materials contained in any third party website referenced in this work.

Contents

How to be a Scientist   2

1 Life Processes   4
What group does this animal belong to?   6
Staying alive   12
Is this living or non-living?   20
How can we help plants to grow?   22
What we have learned about life processes   26

26 − 4 − 1 = 21 − 11 [10]

2 Materials   28
Materials around us   30
Is it magnetic?   36
Just right for the job   40
Sorting materials   46
What we have learned about materials   48

48 − 28 − 1 = 19 − 11 = [8]

3 Flowering Plants: 50
Parts of a flowering plant   52
What do plants need so they can grow?   54
How do plants take in water?   60
Healthy plants   64
Not too hot and not too cold!   68
What we have learned about flowering plants   70

70 − 50 − 1 = 19 − 11 = [8]

4 Introducing Forces   72
Pushes and pulls   74
Making shapes with forces   78
Forces can stop things moving   80
Forces can affect speed and direction   84
What we have learned about introducing forces   88

88 − 72 − 1 = 15 − 11 = [4]

5 The Senses   90
Touch   92
Taste   98
Smell   102
Sight   106
Hearing   110
What we have learned about the senses   114

114 − 90 − 1 = 23 − 11 = [12]

6 Keeping Healthy   116
The life processes   118
Diet and exercise   120
Damaging foods   130
What we have learned about keeping healthy   134

134 − 116 − 1 = 17 − 11 = [6]

Glossary   136

[48]

How to be a Scientist

Scientists wonder how things work. They try to find out about the world around them. They do this by using scientific enquiry.

The diagram shows the important ideas about scientific enquiry.

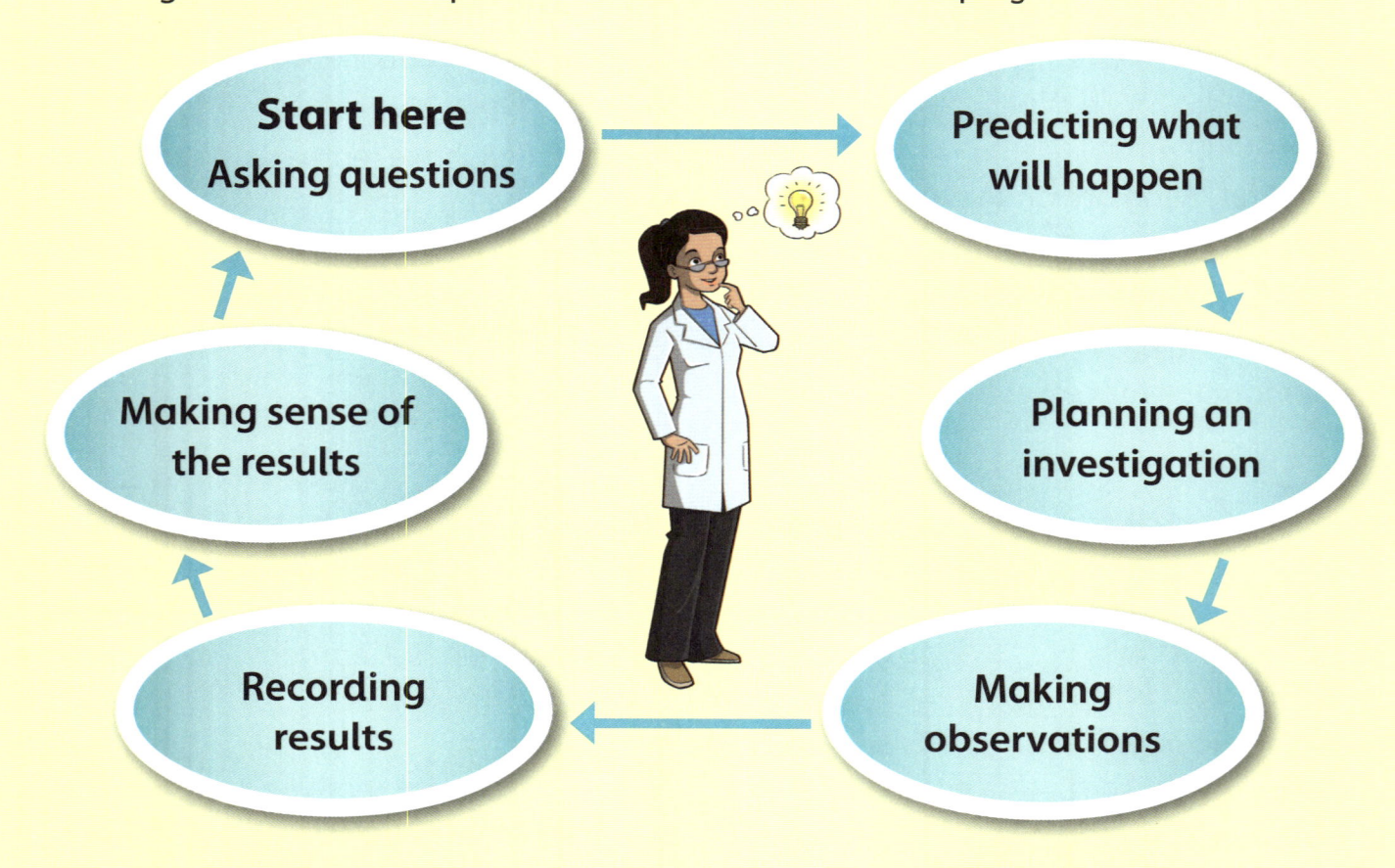

- **Start here** Asking questions
- Predicting what will happen
- Planning an investigation
- Making observations
- Recording results
- Making sense of the results

An example investigation:

Do plants need water to grow?

- Does the plant still live?

Asking questions

Start your questions with words like 'Which', 'What', 'Do' and 'Does'.

- What will happen if we do not give water to plants?

Predicting what will happen

A prediction is when you say what you think will happen in your investigation.

A prediction is more than a guess. Use what you know about plants to help you.

Here is an example of a question and a prediction.

Question

How can we prove that plants need water to grow?

Prediction

Yes. Plants need water to stay alive.

Planning an investigation

When you plan an investigation think about how you will make it a fair test.

What will you keep the same?

- The type of plant, the amount of light and soil, the amount of time the plant grows.

What will you change?

- The amount of water

The things you keep the same or change are called variables.

Making observations

You will measure time.

You will test to find out which plants are the healthiest.

You will look at and measure the height of the plants.

Which plants are tall and green?

Which plants look smaller and brown?

Recording results

There are many ways to record results. You have to think about the best way for your investigation.

A good way is to complete a table. You can also use your results to make a chart or graph.

A table keeps all of your results neat and tidy. A chart or graph can help you to see patterns.

Making sense of the results

At the end of your investigation you must look at your table carefully. You are comparing the plants.

You then decide if the plants needed water. Did the plants without water die? Is this what you expected?

Was your prediction correct?

1 Life Processes

In this module you will:

- learn how we can sort living things into groups
- find out how humans and animals stay alive
- understand the difference between living and non-living things
- know how to keep plants alive.

Word Cloud: reproduce, grow, animal, sort, feed, group, move, senses

These dolphins are one type of animal. The dolphins are alive and healthy because they are able to eat, drink, move, grow and reproduce. These are the life processes that stop animals from dying.

💬 Look at these animals.

| chimpanzee | rat | gorilla | mouse |

Compare the animals. Talk about what is similar and what is different.

Think about what they look like, where they live, what they eat, how they move, etc. These are the animals' features.

Biology

Life Processes

Amazing fact

There are at least six million kinds of animal in the world making up the animal kingdom!

Think about...
Would sorting the animals into groups help us?

5

What group does this animal belong to?

Learn how we can sort living things into groups.

The Big Idea

We can put animals that are like each other into groups.

Scientists have sorted **animals** into **groups** to make it easier to learn about them. To do this, the scientists look at the features of the animals, like you did when you compared the mouse, rat, chimpanzee and gorilla.

parrot

rabbit

eagle

camel

chicken

6

Look at the animals on page 6.

✏️ Can you **sort** the animals into two groups? Write the names of the animals in the table.

There are five animals, so the groups will not be equal.

Group 1	Group 2

✏️ Explain why you put the animals into these groups. Is there a feature that one animal has that another does not have? Do some animals have the same feature?

One way that scientists group animals is to look at the animal's bone structure. Some animals, like hippos, have a skeleton inside their bodies. Other animals, like crabs, have a skeleton (shell) outside their bodies and some, like jellyfish, have no skeleton at all.

The animals with backbones inside them are called vertebrates.

Vertebrates are the group we belong to. Only 10 per cent of the animals that live on Earth are vertebrates.

That means that 90 per cent of the animals that live on Earth are animals with a skeleton outside their bodies or with no skeleton at all. These animals are called invertebrates.

✏️ Can you sort the animals in the word bank into vertebrates and invertebrates? Write them in the table.

Vertebrates	Invertebrates

Word Bank
camel   butterfly   worm   rabbit
spider   horse   bird   snail

Think about...
How can we sort a group of animals, like the vertebrates, into smaller groups?

Life Processes

7

What group does this animal belong to?

Learn how we can sort living things into groups.

The Big Idea

We can sort animals that are like each other into smaller groups.

Only 10 per cent of the animals on Earth are vertebrates, but they are still a very large group of animals to study. So it is useful for scientists to divide vertebrates into smaller groups.

The diagram shows how we can sort vertebrates into five smaller groups: mammals, amphibians, birds, reptiles and fish.

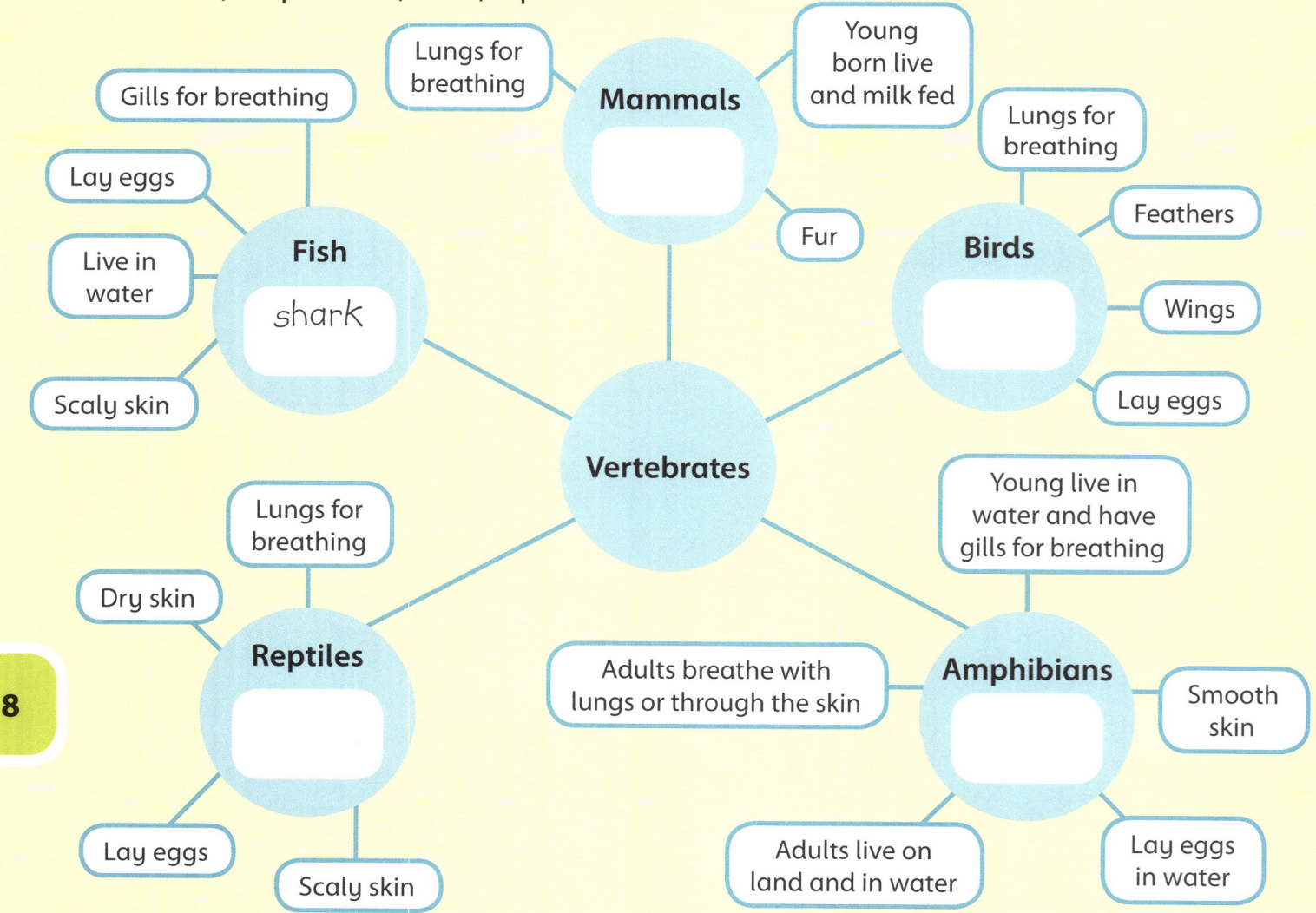

Mammals
- Lungs for breathing
- Young born live and milk fed
- Fur

Birds
- Lungs for breathing
- Feathers
- Wings
- Lay eggs

Fish: shark
- Gills for breathing
- Lay eggs
- Live in water
- Scaly skin

Reptiles
- Lungs for breathing
- Dry skin
- Lay eggs
- Scaly skin

Amphibians
- Young live in water and have gills for breathing
- Smooth skin
- Adults breathe with lungs or through the skin
- Adults live on land and in water
- Lay eggs in water

✏️ Sort the animals in the word bank into the correct spaces on the diagram. One has been done for you.

Word Bank

goat   trout   camel   snake   ~~shark~~   lizard   toad   sparrow   ostrich   frog

✏️ Look at the diagram on page 8 again and complete the sentences.

F_____ and young amphibians have g_____.
Reptiles, f_____, b_____ and amphibians lay e_____.
Fish and r_____ have s_____ skin.

✏️ Look at the photos of vertebrates. Underneath each photo write which group (mammal, reptile, amphibian, bird or fish) it belongs to and why it belongs to that group.

Think about...
Which group does a whale belong to?

1 2 3

Life Processes

9

What group does this animal belong to?

Learn how we can sort living things into groups.

The Big Idea

Some animals are more difficult to sort.

✏️ Think back to the last unit and answer the questions.

1 What are the five groups of vertebrates?

2 Which group do humans belong to?

3 Name two features that mammals have, but other vertebrates do not have.

Some animals might look as if they belong to one group but, watch out, look closely and you will have to think again!

> I live in the sea all my life but I do not have gills to breathe. I have lungs so I must come to the surface of the sea to breathe. I **feed** my young milk. Which group do I belong to?

I live in the sea for half my life and can dive half a kilometre under the sea, but I do not have gills to breathe. I have lungs so I must come to the surface to breathe. I lay eggs and have feathers. Which group do I belong to?

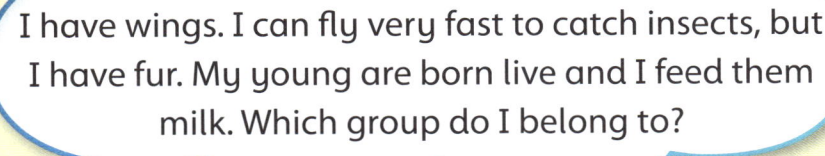

I have wings. I can fly very fast to catch insects, but I have fur. My young are born live and I feed them milk. Which group do I belong to?

Think about...
Can you think of ways of making even smaller groups of animals?

Which groups do these animals belong to?

Answer the questions about how animals are grouped together.

1 What are the two main groups of the animal kingdom?

2 In the vertebrate group there are five groups. Humans belong to a group called mammals. What are the other four groups called?

3 Which animal group does the ostrich belong to? Why?

Now turn to page 26 to review and reflect on what you have learned.

Staying alive

Find out how humans and animals stay alive.

The Big Idea

Animals need to stay alive. How do they do it?

💬 Why are penguins birds? Why are whales mammals? Think of two reasons for each animal.

Like humans, all animals need to be able to breathe and have food to eat and water to drink to stay alive. These are the essential life processes.

💬 This lion is chasing antelope. Is it doing this for fun or for its dinner?

To get food and drink animals need to be able to **move**. Eating food and drinking makes them **grow** and mature, so they can **reproduce** and have young. Moving, growing and reproducing are also life processes.

💬 This person is running. Is she running to catch food? She might be running to the supermarket to buy food!

Amazing fact

The cheetah is the fastest land mammal. It can run at speeds of about 100 km per hour but only for short lengths of time. The cheetah prefers to creep up on its prey then at the last minute run very fast. It can go from standing to 96 km per hour in 3 seconds so its prey does not have much chance to escape.

✏️ Draw a line to match each animal with the food it eats.

| eggs | grass | fish |

✏️ In the box below each animal write one word to describe how it moves.

✏️ Look at the photo of a market. How many foods can you see? Write the names of the foods in your Investigation notebook.

✋ **Investigation: Fruit and vegetables**

Next time you go food shopping with your parents count all the different types of fruit and vegetables. Try to find out out where they came from.

Think about...
Why do cats hunt at night?

Life Processes

13

Staying alive

Find out how humans and animals stay alive.

The Big Idea

Animals use their **senses** to stay alive.

We use our senses to stay alive. Most animals have much better senses than we have. That is because they need their senses to hunt or to help them escape when they are being hunted.

✏️ Draw a line from the body part to the correct sense. One has been done for you.

Eyes	taste
Ears	smell
Nose	touch
Hands	see
Tongue/mouth	hear

✏️ Look at the photo. Which three senses are the lion and antelope using the most in this chase to stay alive?

To find food and water you need to see or smell where it is.

To make sure you are not eaten you need to see and hear your attacker before they get close to you.

Amazing fact

A cat uses its senses to protect itself and to hunt.
- Cats can see much better than we can.
- Cats can sense smell much better than we can.
- Cats can hear sounds a lot better than we can. They can move their ears in many different directions.

Investigation: Where did the sound come from?

1. Sit in a circle with your group.
2. Take it in turns to sit in the middle of the circle blindfolded.
3. One person in the circle drops a marble onto a metal tray.
4. Ask the blindfolded student to point in the direction of the sound.
5. Give the tray and the marble to a different student.

Repeat the activity five times.

A cat is right every time. How about you?

Draw an imaginary animal that is good at staying alive in the wild.

Draw your animal in your Investigation notebook. How does the animal move? What senses does it need? For example, does it need very big ears?

Show your drawing and talk about why your animal will survive well in the wild.

Life Processes

15

Think about...
Why is it important that animals have young?

Staying alive

Find out how humans and animals stay alive.

The Big Idea

Animals grow up, mature and reproduce.

Having enough food to eat allows our bodies to grow and mature. This baby will grow up and may have babies of its own one day.

six months old

These photos show the same person, as he grows from a baby into an adult. Estimate the person's age in each photo and then write the ages on an 'age line', from 0 years old to 40 years old.

16

What is your height?

If possible, measure the heights of a baby, a toddler, a child, a teenager and an adult. Record their heights on the bar chart. Record your height too. What do you notice?

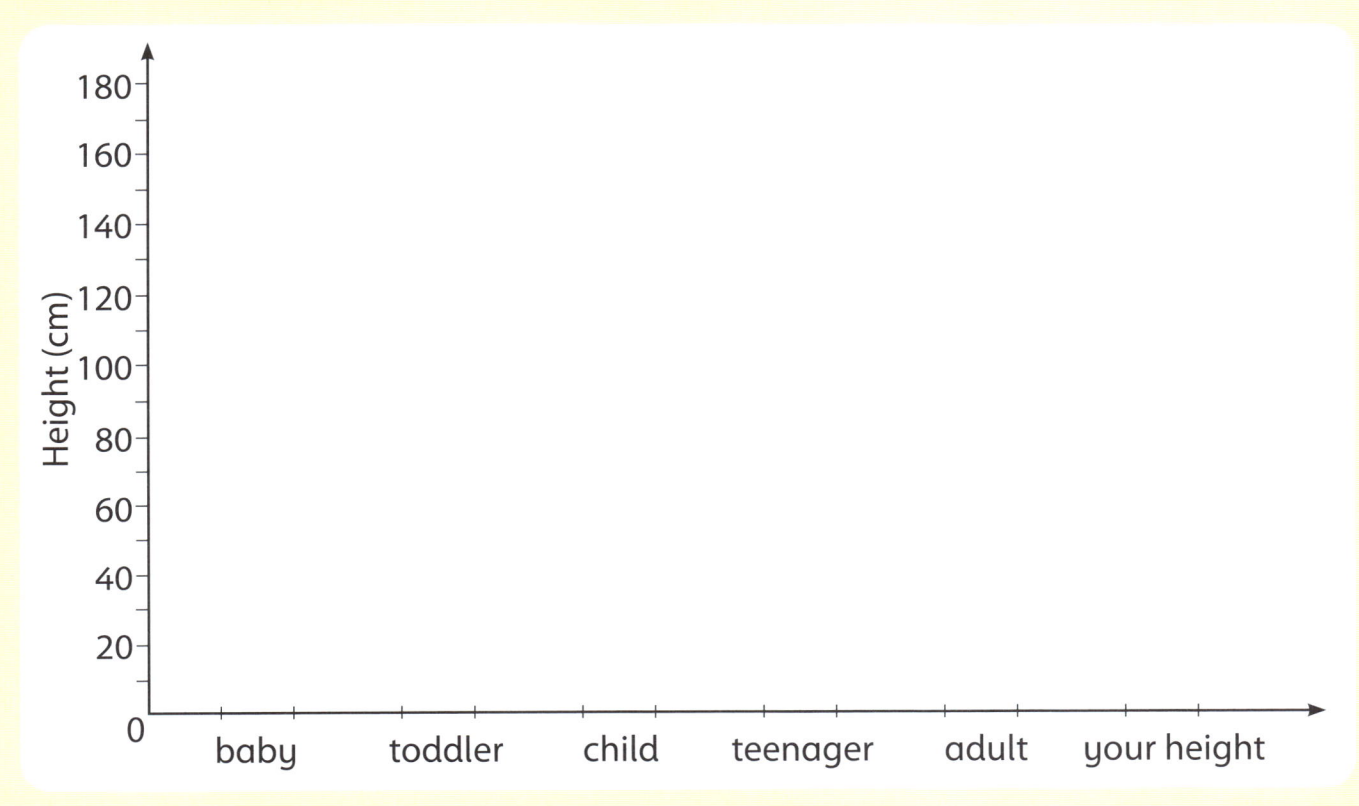

[Bar chart with y-axis labelled "Height (cm)" marked 0 to 180 in increments of 20, and x-axis categories: baby, toddler, child, teenager, adult, your height]

Humans grow up to become adults. Adults are able to reproduce and have families of their own. All animals grow and reproduce.

Growing and reproducing are very important life processes. If animals did not give birth to babies they would eventually die out or become extinct.

Arabian oryx

Amazing fact

The Arabian oryx became extinct in the wild in 1972 in the Middle East. In 1982, thanks to the efforts of many people, including King Khalid Abdul Aziz, it was re-introduced to the wild. Now there are over 1000 Arabian oryx.

Life Processes

Staying alive

Find out how humans and animals stay alive.

The Big Idea

Animals look after their young in different ways.

Animals grow up and reproduce. They have offspring so that their species does not die out.

💬 How many adults and how many children are there in your family?

Some animals, like humans, produce only a few offspring. These animals care for their offspring while they are growing up.

Other animals, like fish, produce lots of offspring. These animals leave their offspring to grow up on their own.

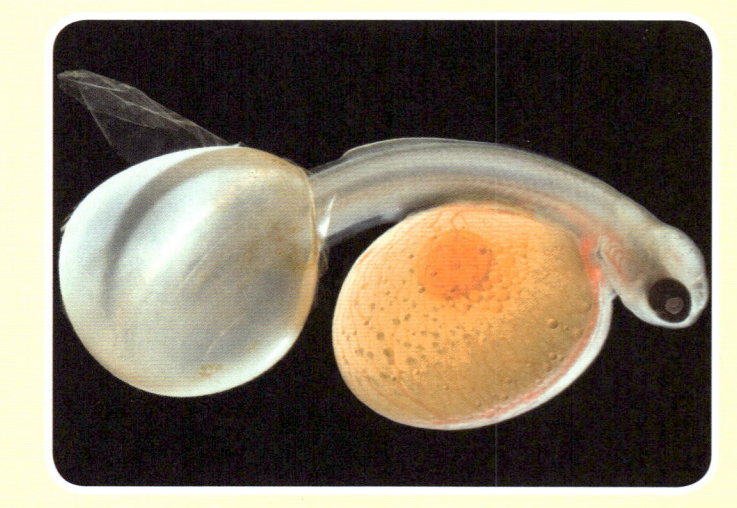

✏️ The following types of animals produce different numbers of eggs.

Animal group	Name of animal	Number of eggs
Birds	Robin	4–6
Reptiles	Crocodile	20–100
Amphibians	Frog	500–1000
Fish	Herring	20 000–40 000

1 Which animal lays the fewest eggs?

2 Which animal lays the most eggs?

3 Describe what is happening to the numbers of eggs in the last column.

💬 Can you explain what the robin can do that the herring cannot do?

✏️ Complete the sentences using the words in the word bank.

All animals stay alive by __breathing__ and _____ food and _____ water.

Animals use their eyes and ears to _____ danger. They _____ quickly to escape.

Young animals _____ to become adults and live for one reason: that reason is to _____.

Word Bank

reproduce   sense   ~~breathing~~
move   drinking   grow   eating

Amazing fact
Twelve rabbits were introduced into Australia in 1859. In less than 100 years there were over 6 000 000 (six million)!

Now turn to page 26 to review and reflect on what you have learned.

Is this living or non-living?

Understand the difference between living and non-living things.

The Big Idea

There are living and non-living things.

✏️ Think about what animals need to do so that they can stay alive. Write three examples.

Animals are living things because they breathe, eat, move, grow and reproduce.

Plants are living things because they grow, move and reproduce. They do not need to move to get food because they make food using energy from the Sun. However, they do move by turning towards the light.

When plants and animals die they are no longer living but they once lived.

A car moves but has never breathed, eaten, grown or reproduced so it is non-living.

✏️ Which of the things in the word bank are living and which are non-living? Write them in the table. One has been done for you.

Living	Non-living
rabbit	

Word Bank

~~rabbit~~   bird   tree   cut flowers
plastic flower   wooden chair
metal bar   concrete   brick   glass

Think about...

Most mammals live on the land. Do you know any that live in water or live in the air? How do they move?

Movement is an important life process. All living things move. Some non-living things move but they cannot do this on their own. A car has to have an engine to move and a person has to start that engine.

Look at these animals.

✏️ Complete the table. Use the words in the word bank. You can use each word more than once.

Group	Movements
Mammals	
Birds	fly, walk, run, swim
Reptiles	
Amphibians	
Fish	

Word Bank

run   walk   jump   swim   crawl   slide   fly

Animals move in different ways on land, in water and in the air.

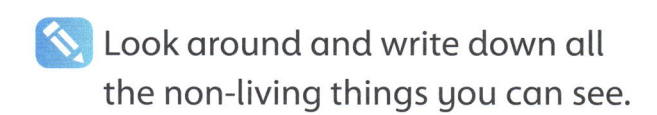

💬 Why do animals move in different ways?

✏️ Look around and write down all the non-living things you can see.

💬 How does each animal move? Can you describe it to a partner? You can use words like creep, crawl, slither or jump.

Now turn to page 27 to review and reflect on what you have learned.

How can we help plants to grow?

Know how to keep plants alive.

The Big Idea

Plants are living things and they need to stay healthy to survive.

Garlic

Centuria

Sunflower

Crocus

✏️ Write three examples of plants that you have seen today. Did you see any of the plants in the photos? Did the plants look healthy?

💬 Look at the picture above. Would you ever see this happening?

Plants have similar living processes to animals but there are differences.

How plants are similar to animals	How plants are different to animals
Plants need food and water.	Plants do not move from place to place.
Plants grow and produce seeds and reproduce.	Plants make their own food from water, air and sunlight.

✏️ Complete the sentences using the words in the word bank.

All plants have roots which bring up ___water___ from the ground.

All plants have a _____ which holds the plant up and delivers water to the leaves and flowers.

All plants have _____ which make food using water and sunlight.

All plants have _____ which attract insects and produce the seeds.

Word Bank

green leaves   stem   flowers   ~~water~~

Life Processes

23

How can we help plants to grow?

Know how to keep plants alive.

The Big Idea

Plants need water and light to stay healthy.

💬 Discuss the four different parts of a plant. What does each part do?

✋ Investigation: Plants in nature

1 Look around the local environment. Look for plants that are growing well.

2 Make some cards like the one below to record the information about your healthy plants. Or copy the card in your Investigation notebook.

3 Repeat the investigation for plants that are not growing well.

💬 Discuss what you found out. What do plants seem to need to grow well?

You are now going to carry out an investigation to find out what plants need to grow well.

Name of the plant:
_____

Place where it is growing:
_____

Amount of sunlight

Low   Medium   High

Amount of water

Low   Medium   High

Height of the plant (cm)
_____

Width of the plant (cm)
_____

Picture or photo of the plant

✋ **Investigation: What does a plant need to grow well?**

You will need four young potted plants.

A Keep one plant in a warm, sunny place and water it regularly.

   Warm + light + water

B Keep one plant in a colder place with light and water it regularly.

   Cold + light + water

C Keep one plant in a warm, sunny place but do not water it.

   Warm + light − water

D Keep one plant covered with black plastic and water it regularly.

   Warm − light + water

1 Watch the plants grow for four weeks.
2 Record the changes each week. Take photographs if you can.

✏️ Draw how the plants looked in your Investigation notebook.

Label each plant as shown below.

Plant A	Plant B
Sun and water	In cold and light

Plant C	Plant D
No water	In dark

💬 Discuss what you have found with your teacher.

Think about...
What would happen to us if the Sun stopped shining on the Earth?

✏️ Write about where you would put a plant in your house to give it the best chance to stay alive. Explain how you would look after the plant.

Life Processes

25

Now turn to page 27 to review and reflect on what you have learned.

What we have learned about life processes

What group does this animal belong to?

✏️ Name one difference between a mammal and a reptile.

✏️ Sort the animals in the word bank into the correct places in the table.

Invertebrates	Vertebrates

Word Bank

crab   deer   eagle   worm
snail   monkey   centipede

I know the difference between a vertebrate and an invertebrate. ○

I can name at least three animal groups. ○

Staying alive

✏️ Imagine you are alone in the desert. What do you need to stay alive? Circle the correct words.

food   clothes   soap   water   books

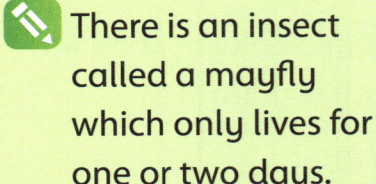

✏️ There is an insect called a mayfly which only lives for one or two days. What is the one life process it must do before it dies?

I can name the life processes. ○

I know the five senses. ○

I understand why reproduction is so important for the survival of animals. ○

Is this living or non-living?

✏️ Is a piece of limestone a living or non-living thing? Explain why.

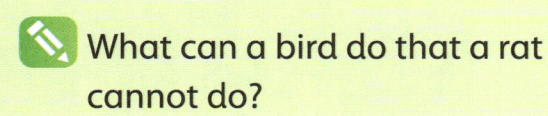

✏️ What can a bird do that a rat cannot do?

I know the difference between living and non-living things. ◯

I know that animals move in different ways. ◯

How can we help plants to grow?

✏️ Unjumble the letters to find the names of the main parts of a plant.

t s m e   _ _ _ _

f e l a   _ _ _ _

w e f o r l   _ _ _ _ _ _ _

o o r t s   _ _ _ _ _

✏️ Name two things that plants need to grow well.

I know the names of the main parts of a plant. ◯

Life Processes

27

2 Materials

In this module you will:
- learn that every material has properties
- find out that some materials are magnetic, but many are not
- explore why different materials are used for different purposes
- learn how to sort materials into groups.

Amazing fact

Some materials can attract other materials.

Think about...

Do you know what this object is called? Can every material do this?

Materials are all around us. All of the objects in the photograph are made of materials.

💬 Find three shiny materials.

Find two materials that have to be strong.

Find two materials that have to be hard.

Find two materials that have to be soft.

These objects are made of glass.

💬 Why is glass such a useful material for a jar?

What would happen if you used a hammer made from glass?

Word Cloud

material
absorbent
object
soft
hard
waterproof
magnetic
sort
group

Materials around us

Learn that every material has properties.

The Big Idea

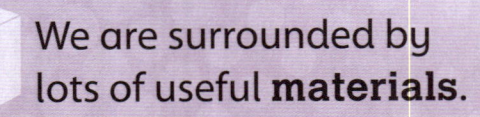

We are surrounded by lots of useful **materials**.

💬 How many useful materials can you see around you?

All the **objects** we use are made of materials. Even our bodies are made of different materials. We can identify the different materials because they look and feel different.

How a material looks and feels is called its properties.

✋ Investigation: What is in the box?

Work with a partner. You are going to investigate different properties. Think about all the properties you know. Use these to help you to identify the mystery objects.

1. Place an object in the mystery box. Do not let your partner see the object.
2. Ask your partner to put their hand in the box and try to identify the object and the material.
3. Take turns to guess the object.

💬 Which objects were difficult to identify?

Which object was the easiest to identify?

Can you match each material to its properties? One has been done for you.

a metal bowl

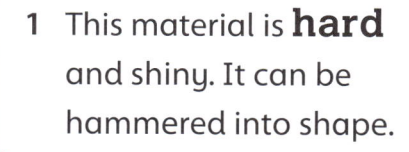

b ceramic vase

c glass bottle

d rubber ball

e wool rug

f stone for building

1 This material is **hard** and shiny. It can be hammered into shape.

2 This material is see-through. It breaks easily.

3 This material is **soft** and feels hairy.

4 This material is shiny and hard. It is not see-through. It breaks easily.

5 This material is hard and rough. It is heavy and strong.

6 This material is soft and bouncy.

Materials around us

Learn that every material has properties.

The Big Idea

We use some materials for special jobs.

✏️ Can you name an object that is made of each material?

Rubber _____   Stone _____

Metal _____   Glass _____

Word Bank
coin   house   ball   bottle

Metals are very useful. People have used metals for thousands of years.

💬 What are the uses of metals in the photographs?

Discuss with a partner all the uses of metals you can think of.

Try to identify which metal is being used for each job. Why is this the best metal for the job?

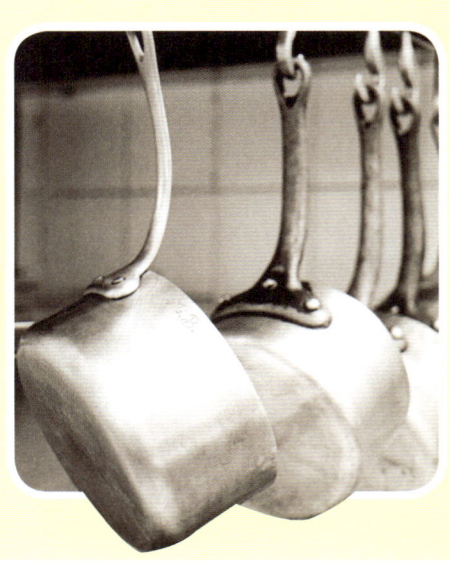

Some metals stay shiny for thousands of years. Other metals react with water and air and quickly change into dull materials. For example, iron can rust if it gets wets.

💬 Play a materials game with a partner.

Student A: Say the name of an object.

Student B: Tell A the best material to make the object from.

> Metal is the best material for a spoon.

Student A: Ask B which is the worst material to make it from.

> Paper is the worst material for a spoon.

Continue until you have both chosen four objects each.

Think about...
Why is gold a popular metal for jewellery?

A gold car would last longer than a steel car. Why don't we use gold to make cars?

✏️ Look at these words. Draw a circle around all the properties of materials.

smooth
(hard)   measure
see-through
strong
happy   metal
rough
shiny   bendy
dull
heavy
soft
copper
observe   ball
paper
magnetic

Materials

33

Materials around us

Learn that every material has properties.

The Big Idea

Materials have different properties. This helps them to do different jobs.

💬 Think about the properties of materials. With your partner, how many different properties can you name? Write a list.

✋ **Investigation: Materials in the school**

We use materials for different things because of their properties.

1 Walk around the classroom or your school. Try to find different materials.
2 Note where you find each material.
3 Test each material to find out what properties it has.

This type of scientific work is called a survey.

💬 How will your group record your results?

Where will you look for materials?

Which properties will you look for?

✏️ Which material am I?

1 I am shiny, hard and very strong. I can be shaped to make many different objects.

2 I am dull and soft. I am white and I am easy to see on black surfaces.

3 I am hard and shiny. You can see through me. Be careful, I break easily.

4 I am bendy and stretchy. I can have different colours.

Glass

Steel

Rubber

Chalk

34

Slippery or grippy?

Shiny floors can be very slippery. This means that people might slip and fall over.

Think about the floors in the school. Which floors are shiny?

Think about...
How can you safely test floors to find out which are slippery and which are not?

Which floors in your school need a 'Do not run!' sign?

What might happen if shiny floors get wet?

Look at the sign at a swimming pool. How does this sign help you to be safe?

Read the questions and circle the correct words.

1 Which of these are properties of materials? (four words)

| metal | hard | soft | wood | shiny | heavy |

2 Which of these are metals? (three words)

| gold | plastic | wood | glass | iron | copper |

3 Which is the best material for making a bell? (one word)

| wood | plastic | paper | metal | ceramic | glass |

4 Which is the best material for making a car windscreen? (one word)

| wood | plastic | paper | metal | ceramic | glass |

Now turn to page 48 to review and reflect on what you have learned

Is it magnetic?

Find out that some materials are magnetic, but many are not.

The Big Idea

Magnets can pull materials towards them without touching them.

Look at the photograph.

💬 What is happening? Can you describe it to your partner?

Some metals have a very important property. They can be made into magnets. Magnets pull some materials towards them. We say that they attract other materials.

Some materials are natural magnets. A rock called lodestone is a magnet. Other magnets have to be made. The first photograph on this page shows a man-made magnet.

A man-made magnet

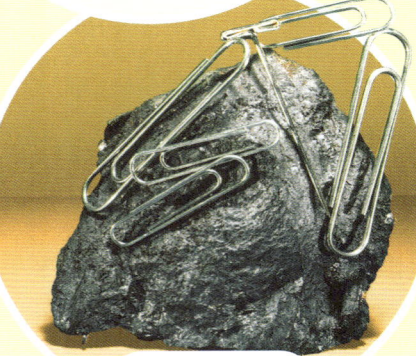

A lodestone

✋ Investigation: Are all materials **magnetic**?

1 Use a bar magnet to test objects.

- If an object is attracted towards the magnet, the object is made of a magnetic material.
- If an object is not attracted towards the magnet, the object is made of a non-magnetic material.

2 Copy the table in your Investigation notebook. Record your results in the table.

💬 Can you see a pattern in your results?

Object	Material	Magnetic or non-magnetic?
1 pencil	wood	non-magnetic
2		

You found out that non-metals are non-magnetic. Some non-metals are wood, plastic and ceramic.

Some metals are magnetic. Iron, steel, nickel and cobalt are magnetic.

Other metals are non-magnetic. Some non-magnetic metals are gold, silver, copper and aluminium.

✏️ Name two magnetic materials.

✏️ Name two non-magnetic materials.

One end of a magnet is the North seeking pole. The other end is the South seeking pole. If the poles pull together. they are attracted. If the poles push apart, they are repelled.

✋ **Investigation: What happens when we bring two magnets together?**

1 Slowly push two magnets together with the North seeking poles facing each other.

✏️ What do you feel?

2 Push the magnets together with the South seeking poles facing each other.

✏️ What do you feel?

3 Push the magnets together with a South seeking pole facing a North seeking pole.

✏️ What do you feel?

You have discovered a scientific law. This is the law of magnetism.

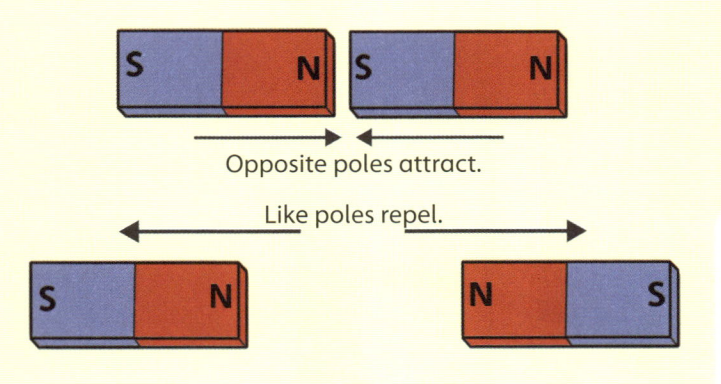

Opposite poles attract.

Like poles repel.

Think about...
The North seeking pole always tries to point North. How can this help you to find your way?

Materials

37

Is it magnetic?

Find out that some materials are magnetic, but many are not.

The Big Idea

Magnets are used in lots of toys and machines.

The four compass points are North, South, East and West.

A compass is an important use of magnets.

✏️ What is the magnet in the compass being used for?

Some amazing uses of magnets

There are many important uses of magnets.

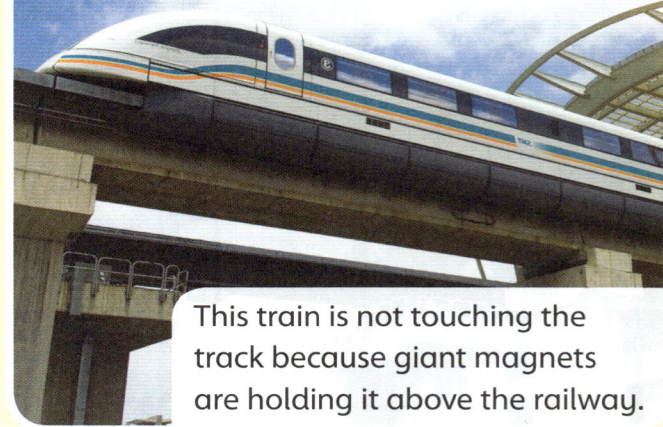

The car is being lifted by a magnet.

This train is not touching the track because giant magnets are holding it above the railway.

You can use magnets to make games and puzzles.

Investigation: Design and make a magnetic game

1. Plan and then make your game.
2. Invite another group to try your game.
3. Put your games on display. Look at all the games carefully.
4. Write the name of the game you think is the best. You cannot choose your own game.
5. Write a reason why you think the game is the best.
6. Your teacher will tell you which game has won the design competition.

Do not worry if you do not win. Some of the greatest inventions in the world did not work at first. How can you make your game better?

Fill in the missing words. One has been done for you.

Some objects __attract__ other objects to them. They are called _____.
A _____ has two ends. One end is the _____ seeking pole and the other end is the _____ seeking pole. The law of magnetism is: Like poles _____ and opposite poles attract.

Iron, _____ and cobalt are magnetic materials. All other _____ and non-metals are non-magnetic. These materials are not _____ to a magnet. There are many uses of magnets.

Word Bank

~~attract~~	attracted	magnet
magnets	metals	nickel
North	repel	South

Materials

39

Now turn to page 48 to review and reflect on what you have learned.

Just right for the job

Explore why materials are used for different purposes.

The Big Idea

A material must have just the right properties to do its job.

Clean up water?

Hammer nails?

Make a mirror?

Make a bell?

Look at the photograph of paper towels.

✏️ Which job do paper towels do best? Why are they not good for the other jobs?

It is very important to choose the best material for the job. For example, building materials need to be strong and glass is used for windows because it is see-through. Now you will investigate a group of materials called fabrics.

A fabric is a cloth made by weaving or knitting. Fabrics are used for carpets, curtains, clothes and coverings for furniture.

40

💬 Which properties are important for the fabric for a T-shirt?

Is it soft? Is it bendy? Does it need to be **waterproof**?

- waterproof?
- hardwearing?
- soft?
- strong?
- lightweight?
- bendy?

✋ Investigation: Which fabric is best for a warm blanket?

Imagine you are choosing a fabric for a warm winter blanket.

💬 How will you test your fabrics to find out:
- which are bendy?
- how strong they are?
- how soft they are?
- which is the warmest?

1 Carry out your tests.
2 Decide which is the best fabric for the blanket.
3 Imagine you work for a shop that sells this fabric. Design a presentation to explain why your fabric is the best.

This person is selling lots of different fabrics.

✏️ Is paper good for making clothes?

1 Write two properties of paper that make it good for making clothes.

2 Write two properties of paper that make it bad for making clothes.

Materials

41

Just right for the job

Explore why materials are used for different purposes.

The Big Idea

We can test the properties of materials to see how useful they are.

💬 Discuss the fabrics you can see in your school.

Which fabrics will last the longest? Which fabrics are the softest? Are there any waterproof fabrics?

We use paper towels for many jobs. The soft paper absorbs water. A good paper towel is very **absorbent**.

Paper towels are very useful for wiping up spills of water.

Not all paper makes good paper towels. It is important to choose the paper with the best properties.

✋ Investigation: Which paper makes the best paper towels?

You are going to test different papers. You need to find the most absorbent paper.

1. Pour water into a plastic bowl.
2. Collect your paper samples.
3. Weigh five pieces of each paper, in grams. 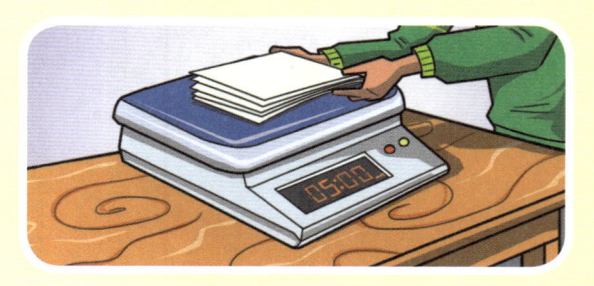
4. Fold up paper sample 1 and put it in the water. 
5. Leave the paper in the water for 1 minute.
6. Take the paper out and let it drip over the bowl for 1 minute. Then weigh the wet paper.

7 Find the weight of the water that the paper has absorbed.

Take the weight of the dry paper away from the weight of the wet paper.

Example: 100 g − 5 g = 95 g

8 Do the test again for each paper sample.

Copy and complete the table in your Investigation notebook. Record all your results.

Paper sample	Weight of five dry paper sheets	Weight of five wet paper sheets	Weight of water absorbed
1	5 g	100 g	95 g

Can you answer these questions about your investigation? Write your answers in your Investigation notebook.

1 Why must each sample of paper be the same size?

2 Why must we cover all of the paper sheets in water?

3 Why do we need to weigh the dry paper sheets?

To test how strong a material is, scientists place heavy weights on it. They add more and more weights until the material breaks.

Think about...

How can you use this method to test your paper samples? How will you find out if paper towels are stronger or weaker when they are wet?

Clue: You can put coins in the middle of your paper towel.

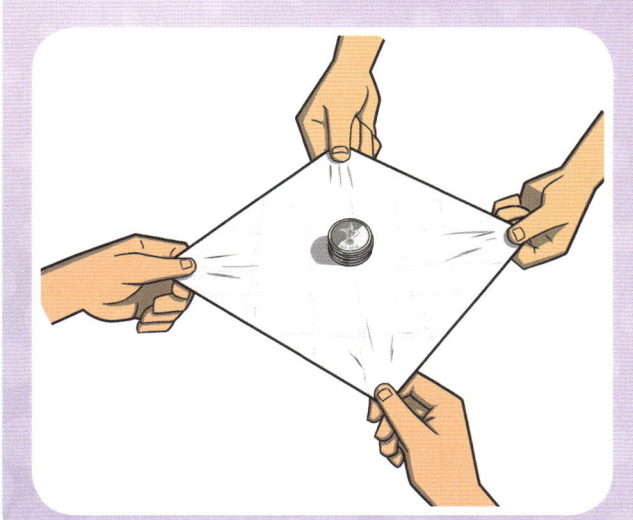

Just right for the job

Explore why materials are used for different purposes.

The Big Idea

Inventors are always trying to find new uses for materials.

Can you match each object with the material it is usually made of?

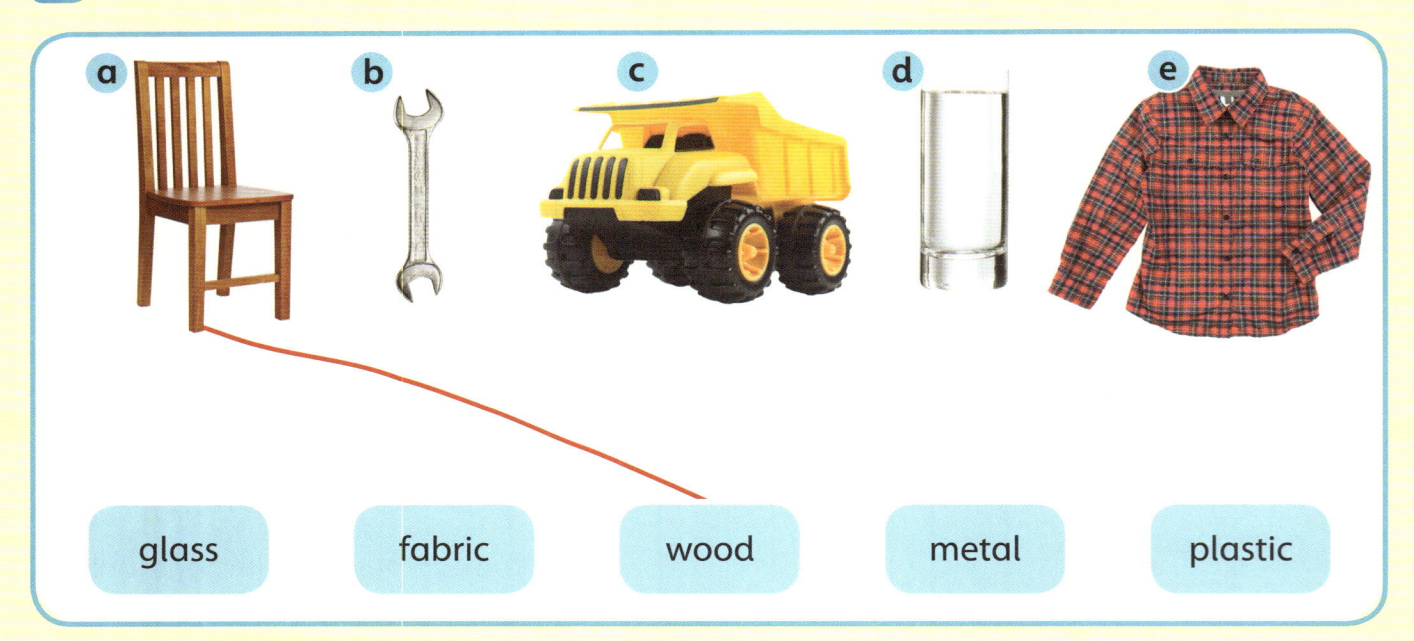

| a | b | c | d | e |

| glass | fabric | wood | metal | plastic |

Invent an object made of a different material.

Your group is going to work as an inventing team. Choose one of the objects in the photographs. You are going to invent a new version of this object. It will be the same shape but made of a different material.

Decide which material you will use. Then find out the properties of this material.

Tell the rest of the class why the object made of your choice of material is better.

Listen carefully when each group shares their ideas with you. Ask them questions. For example:

- Will your object be strong enough?
- Will it break easily?
- Will it be too heavy?
- Will it be too expensive?

Think of some more questions to ask the other groups.

Mixed materials

Some objects are made of more than one material. A screwdriver has a metal tip and a plastic handle. The metal is strong so it can turn the screws. The plastic handle is softer and more comfortable for your hand. The plastic also protects you from electric shock.

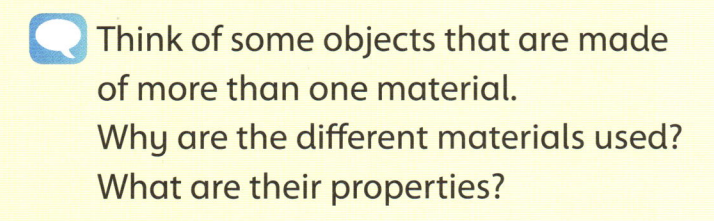

💬 Think of some objects that are made of more than one material.
Why are the different materials used?
What are their properties?

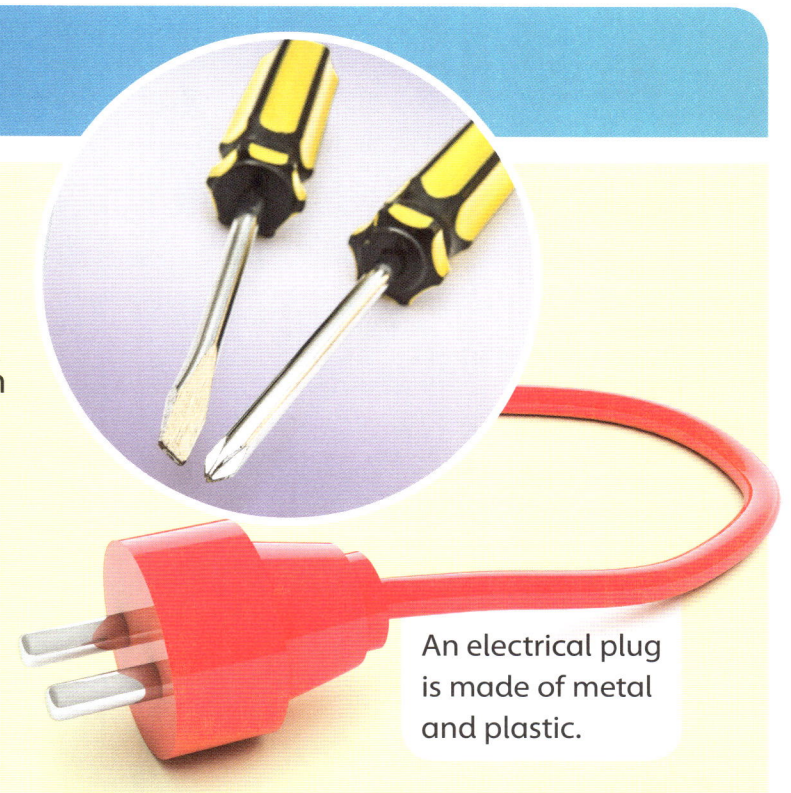

An electrical plug is made of metal and plastic.

✏️ Answer these questions.

Write three properties of metals.

Explain why paper towels must be very absorbent.

Tissue paper is very absorbent. Why is it not used for paper towels?

Explain why a hammer is made of metal but has a wooden or plastic handle.

Now turn to page 49 to review and reflect on what you have learned.

Sorting materials

Learn how to sort materials into groups.

The Big Idea

We can group materials into families depending on their properties.

💬 Look at the hair dryer. Can you see two different materials that are used to make the hair dryer?

What are the metal parts of the hair dryer used for?

Why is it important to have some plastic parts?

✋ **Investigation: Make a display of materials**

You will need a large piece of card and lots of small objects.

You are going to **sort** the objects into **groups** and make a display.

You can sort the objects into different materials – metal, wood, plastic, glass, ceramic and fabric.

Or you can sort the objects into different properties – hard, soft, bendy, rigid, heavy, light, absorbent, waterproof, transparent, opaque, shiny and dull.

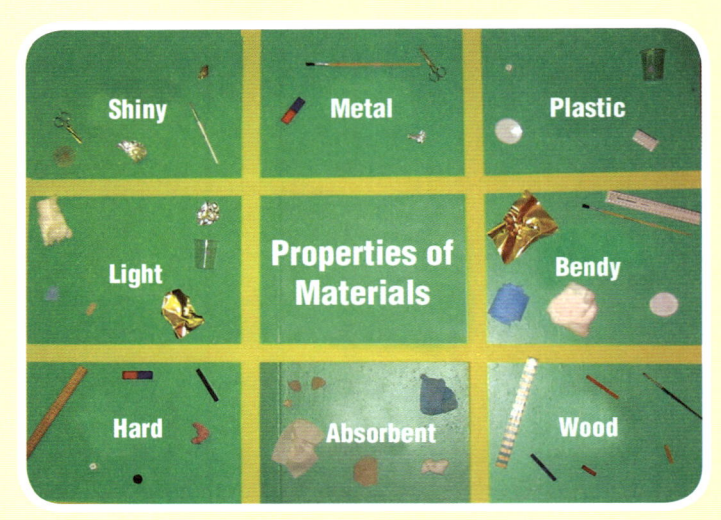

1. Draw lines on your card to divide it into squares. Draw one square for each material or property.

2. Write the name of a material or property in each square.

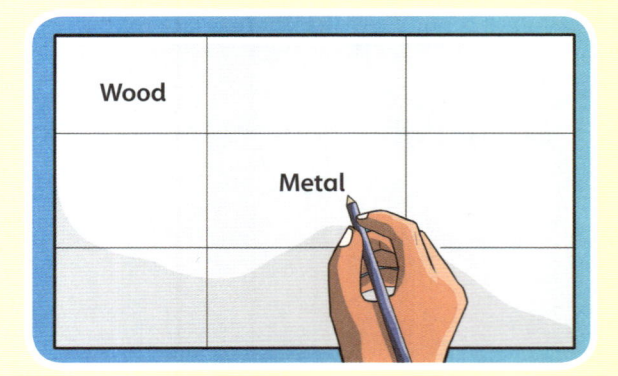

3 Put each object in the correct square.

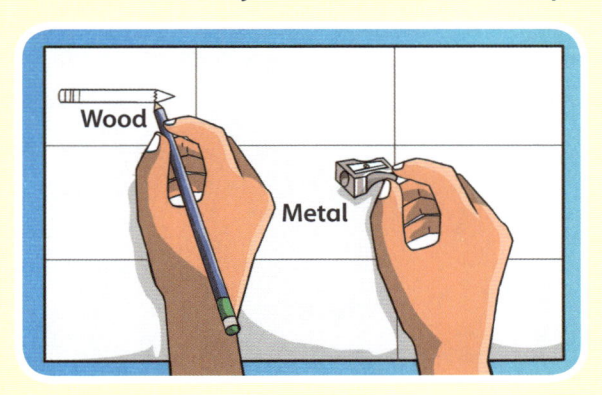

We can split each group of materials into smaller groups. For example, there are hard woods and soft woods. There are magnetic metals and non-magnetic metals.

💬 How many types of plastics do you think there are?

For example, think about a hard light switch and a soft plastic bottle.

✏️ Complete the table of materials.

Material	Some properties	Examples of what it is used for
Metal	hard, strong, shiny, easy to shape	
	transparent, breaks easily, waterproof, shiny	
Wood		furniture, spoons, doors
	hard, strong, waterproof	toys, water bottles, cases for hair dryers
	soft, bendy, absorbent	clothes, curtains, towels

Now turn to page 49 to review and reflect on what you have learned.

What we have learned about materials

Materials around us

✏️ Can you unjumble the six properties of materials? Write the words in the boxes.

Then match each property with the opposite property.

1 f o s t — soft — smooth
2 h r g o u — — rigid
3 n e b y d — — dull
4 h i y s n — — hard
5 a g m e n t c i — — weak
6 g o r n s t — — non-magnetic

(soft is connected to hard)

- I know that different materials have different properties. ○
- I understand why different materials are used for making different objects. ○

Is it magnetic?

✏️ Which of these objects do you think are attracted to a magnet? Circle the correct pictures.

Steel nail | Plastic toy | Gold ring | Aluminium pan | Glass jar | Iron railings

- I know two properties of magnets. ○
- I know which materials are magnetic. ○

48

Just right for the job

✏️ Can you complete the table? Write a good material for each object. Write one property that makes the material right for the job.

Object	Ball	Towel	Cup
Material	Rubber		
Property	Bouncy		

I know why different materials are used to make different objects. ◯

I understand why many objects are made from a mixture of materials. ◯

Sorting materials

We can sort objects into groups according to their properties or the material they are made of.

✏️ Which of these words are properties? Circle the properties in red.
Which of these words are materials? Circle the materials in blue.

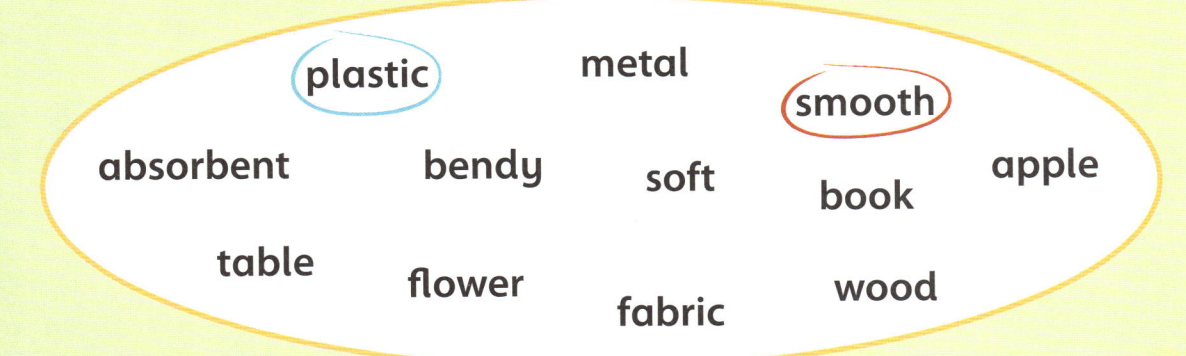

(plastic) metal (smooth)
absorbent   bendy   soft   book   apple
table   flower   fabric   wood

I can sort objects into groups depending on their properties. ◯

I can sort objects into groups depending on the material they are made of. ◯

3 Flowering Plants

In this module you will:

- learn that plants have roots, leaves, stems and flowers
- understand that plants need water and light to grow
- describe how plants take in water through their roots and how the water moves through the stem
- understand that plants need healthy roots, leaves and stems to grow well
- learn that temperature changes the way plants grow.

Amazing fact

Did you know that scientists have identified over 298 000 species of plants?

Biology

💬 Do you recognise the fruit that is growing? Can you see any flowers?

These are flowering plants. Without plants, life on Earth would not exist. Humans and animals need plants for their food.

Word Cloud

root
flower
stem
leaf
water
grow

Flowering Plants

51

Parts of a flowering plant

Learn that plants have roots, leaves, stems and flowers.

The Big Idea

Flowering plants have four main parts.

🗨️ Look at this photograph. What is in the vase?

✏️ Look at the picture of a flowering plant. Label the parts using the words in the word bank.

stem

Word Bank

flower   ~~stem~~   leaf   roots

🗨️ Each part of the flowering plant does different things to keep the plant alive and healthy. What do you think each part does?

Flower

In many plants the **flower** has a nice colour and a nice smell. This attracts insects. The flower is where seeds are produced.

Stem
The **stem** supports the plant. It keeps the plant upright. The stem also transports water and food around the plant. Stems can be flexible or woody.

Leaf
The **leaf** makes food for the plant from sunlight energy. Most of the food the plant needs is made in the leaves. Scientists have a special name for this process, it is called photosynthesis.

Roots
Roots keep the flowering plant anchored in the soil. Some roots have tiny hairs on them to help get **water** from the soil. Roots are very important because plants need water to grow.

✏️ Draw the different parts of your plant in the boxes.

Stem

Roots

Flower

Leaf

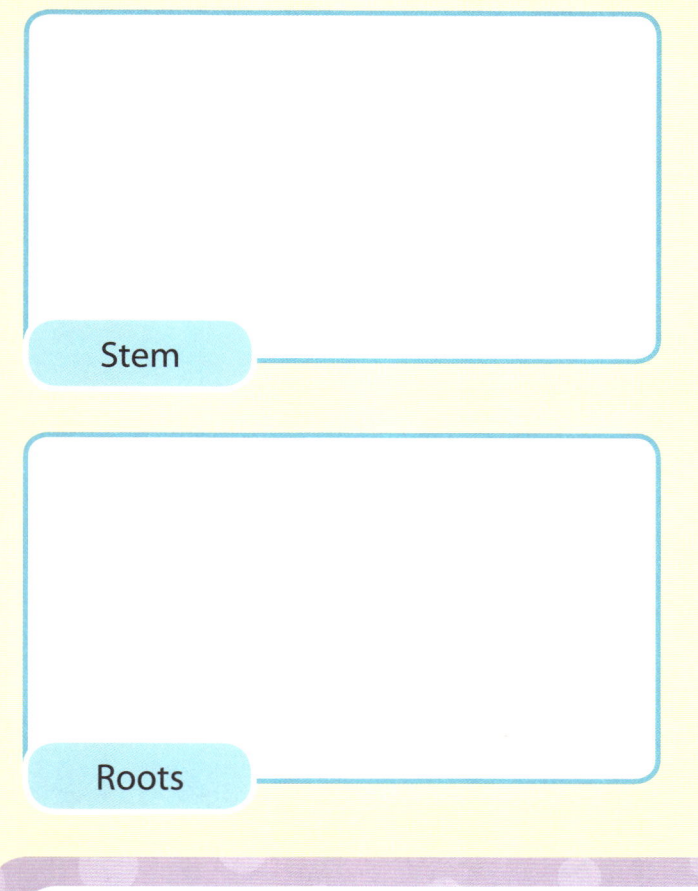

✏️ Find and circle the parts of a flowering plant in the wordsearch.

t	u	r	x	k	m	i
f	s	k	o	f	g	h
c	t	y	c	o	j	s
d	e	a	s	v	t	l
a	m	d	l	o	y	l
l	j	u	b	p	x	e
f	l	o	w	e	r	a
l	z	w	r	h	p	f

Flowering Plants

53

Now turn to page 70 to review and reflect on what you have learned.

What do plants need so they can grow?

Understand that plants need water and light to grow.

The Big Idea

We can carry out investigations to show that plants need water and light.

💬 Look at the photographs. How are the plants getting light and water?

💬 How can we show that plants need water and light to **grow**? What sort of investigations can we do?

54

✏️ Write down two or three ideas about how you will carry out the investigations. This will be your plan for your investigations.

Investigation 1: water

Investigation 2: light

💬 Talk about what you think will happen to the plants when you carry out your investigations.

When we make suggestions about what might happen in an investigation we are making a prediction.

✏️ Write a prediction for what you think will happen to the plants in these investigations.

Investigation 1: water

I predict that

Investigation 2: light

I predict that

💬 How can we make sure that our investigation of plants is a fair test?

When we do investigations we need to make sure that the test is fair. We must make sure that we change only one thing and keep everything else the same.

Think about...
What might happen if a plant has too much sunlight or rain?

55

What do plants need so they can grow?

Understand that plants need water and light to grow.

The Big Idea

Without water a plant will not be able to grow.

💬 What is a fair test? What is a prediction? Why do we need a plan?

✋ Investigation: Do plants need water to grow?

1. In your group choose two plants. Make sure the plants are the same size and same type. This is to make it a fair test.
2. Label one plant 'Water me'. Label the other plant 'Do not water me'.
3. Place your plants on a windowsill.
4. Each day, give a little water to the plant labelled 'Water me'. Give no water to the plant labelled 'Do not water me'.
5. Copy and complete the table in your Investigation notebook. Each day, write down what you see. These are your observations.

Day	Plant that has been watered	Plant that has NOT been watered
1	Plant looks healthy	Plant looks healthy
2		

Once you have collected all your observations, you can use them to decide whether plants need water to grow well. This is called your conclusion.

✏️ Write your conclusion here.

> I think plants

✏️ Draw how your plants looked after 7 days and label them.

✏️ Fill in the missing words using the word bank.

When we want to investigate something we need to make a ____plan____. Investigations often start with a q_____ to test. Our question was 'Do plants need water to grow?' We can make suggestions about what we think will happen, this is called making a p_____. We need to make sure that the investigation is a f_____ test. For this investigation we made sure the p_____ were the same size and type.

Word Bank

question   prediction   plants   ~~plan~~   fair

Flowering Plants

57

What do plants need so they can grow?

Understand that plants need water and light to grow.

The Big Idea

Without light a plant cannot grow.

Do you remember your plan for investigation 2? Look back at page 55.

✋ **Investigation: Do plants need light to grow?**

1 Choose six seedlings. Make sure they are the same size and type.

2 Put three seedlings on a growing dish. Make a label 'Light'. Put the growing dish on a windowsill.

3 Put the other three seedlings on a growing dish. Make a label 'No light'. Put the growing dish in a dark place.

4 Give a little water to both sets of seedlings each day.

5 Copy and complete the table. Each day write your observations.

Day	Seedlings that have been in the light	Seedlings that have NOT been in the light
1	Seedlings look healthy	Seedlings look healthy

When you have collected all your observations, you can use them to decide whether plants need light to grow well.

✏️ Write your conclusion here.

> I think plants

✏️ Draw how your plants looked after 7 days and label them.

💬 Look back at your predictions for the two investigations on page 55. Were they correct?

Why do plants need light to grow?

Look at the picture of the flowering plant. The leaves of a plant use the energy from the Sun in a special way to create food for the plant. This is called photosynthesis. For photosynthesis to happen the plant also needs carbon dioxide from the air and water from the ground.

✏️ The plant needs food. Where does it make this food?

✏️ What three things does the leaf need so that it can make the food? Fill in the labels.

Word Bank

carbon dioxide   sunlight   water

✏️ Unjumble the letters to find some words to do with plants.

1 h o s h s t n y e t i s p o

2 t e s m

3 t e w r a

4 g h l t i   Light

5 t o r o

6 r b o c n a   i x o d e d i

7 f a l e

Now turn to page 70 to review and reflect on what you have learned.

How do plants take in water?

Describe how plants take in water through their roots and how the water moves through the stem.

The Big Idea

Roots have tubes that carry water around the plant.

💬 Why is water important for a plant? How does the plant get its water?

Plants get their water from the soil and they do this through their roots. Some roots have small hairs. The water passes through the root hairs to a 'tube' in the root, which takes the water to the stem and then to the leaves.

Look at this picture showing small root hairs.

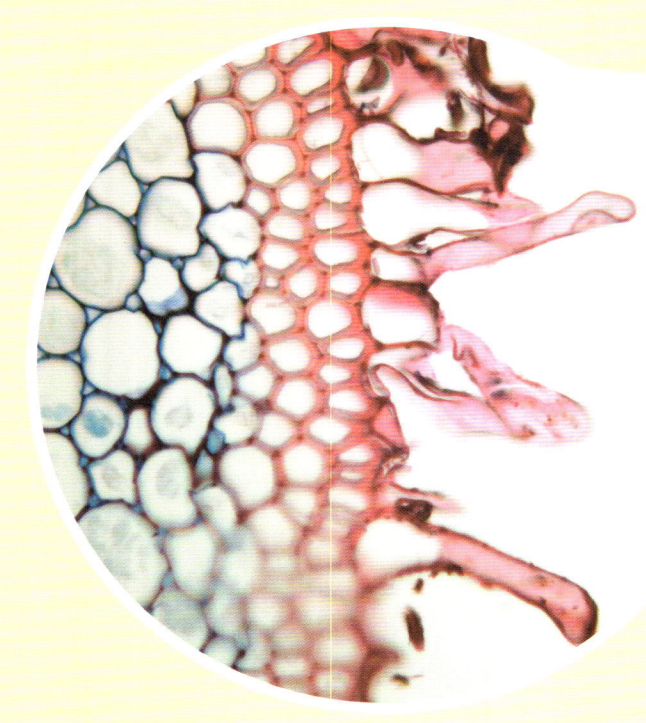

60

✏️ On the diagram draw arrows to show the route the water travels through the plant. One arrow has been done for you. Complete the diagram by labelling the different parts of the plant.

Remember
Can you remember why a plant needs water to reach the leaves? What happens in the leaves? What is this process called?

✏️ Roots are important to plants. Read the list and tick the two things that roots do.

Keep the plant secure in the ground ☐

Make food ☐

Get water from the soil ☐

Give the plant its colour ☐

Get energy from the Sun ☐

Flowering Plants

61

Think about...
Where is the best place to water a plant? At the roots or on the leaves?

How do plants take in water?

Describe how plants take in water through their roots and how the water moves through the stem.

The Big Idea

We can see how water travels through the stem.

💬 How can we investigate how water travels through a stem?

We are going to investigate how water is transported from the root to the stem. We will use celery for our investigation because celery has thick stems.

Look at this picture of an investigation that has been set up.

✋ Investigation: How does a stem transport water?

1 Collect your celery stem.
2 Cut off 2 cm from the bottom of the stem.
3 Pour water into the container until it is half full.
4 Choose a food colouring. Add 10 drops of food colouring to the water.
5 Stir the water and the food colouring very gently with the celery. Make sure that all the food colouring is mixed properly.
6 Put the container in a light place – perhaps on a windowsill.

💬 What do you think will happen to the celery?

✏️ Write down your prediction.

I think that

✏️ How long do you think it will take before we see anything happening?

I think it will take

✏️ Check your investigation over the next 2 days to see if there are any changes. After 2 days colour in the picture to show what has happened to the celery.

✏️ Write a sentence to describe what happened.

Think about...
What has happened to this plant's roots? Do you think it is a healthy plant?

True or false?

Some plants have root hairs.	(True)	False
Water is transported from the stem to the roots.	True	False
We cannot investigate how water is transported through the stem.	True	False
Plants need water to grow.	True	False

Flowering Plants

63

Now turn to page 71 to review and reflect on what you have learned.

Healthy plants

Understand that plants need healthy roots, leaves and stems to grow well.

The Big Idea

Plants need to be healthy to grow well.

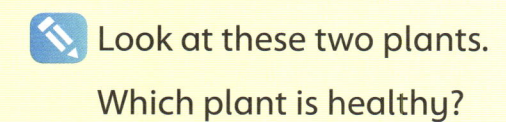

✏️ Look at these two plants.

Which plant is healthy?

Which plant is unhealthy?

We have carried out investigations which have shown us that plants need light and water to grow well. When plants do not have these things they become unhealthy.

💬 How can we help this plant get better?

Roots need to spread out to find water. If they are squashed into a small pot they cannot do this. When roots are squashed in a small pot we say they are pot bound.

✏️ Complete the 'doctor's report' on this plant. Use the word bank to help you.

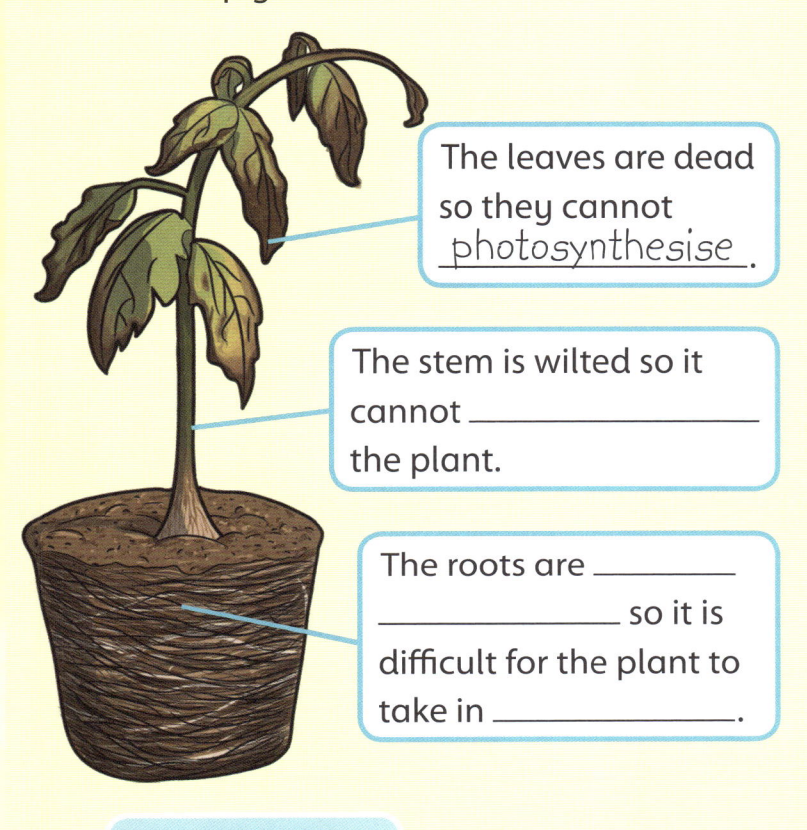

The leaves are dead so they cannot __photosynthesise__.

The stem is wilted so it cannot _____ the plant.

The roots are _____ _____ so it is difficult for the plant to take in _____.

Word Bank

~~photosynthesise~~   water
pot bound   support

✏️ Write a letter to the owner of the plant. Explain what they should do to make their plant better. Write your letter in your Investigation notebook.

Think about...
What do you think happens to a tree when it does not have enough water?

✏️ Look at these flowering plants and plant pots. Match each plant with the best size plant pot. Colour in the plants so they look healthy.

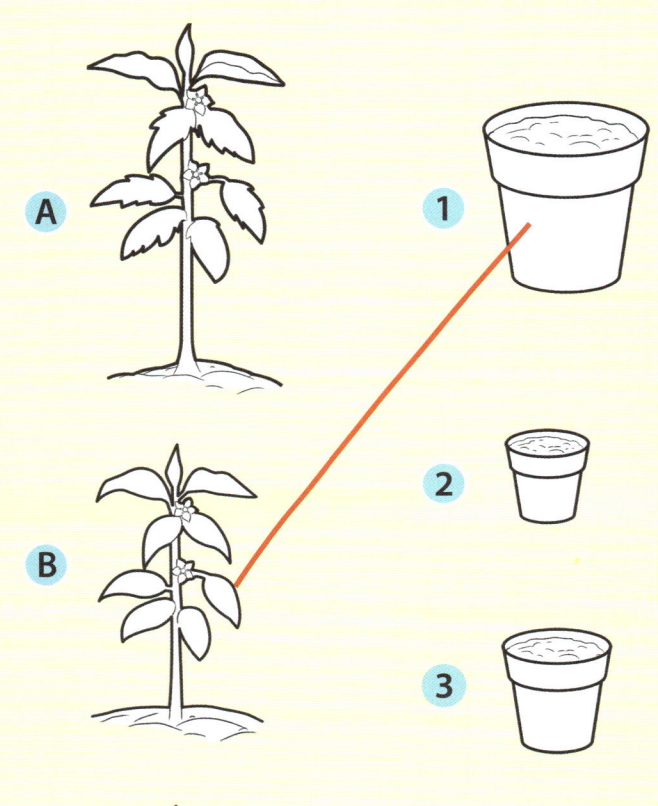

A
B
C
D
E

1
2
3
4
5

Flowering Plants

65

Healthy plants

Understand that plants need healthy roots, leaves and stems to grow well.

The Big Idea

Trees are very large plants. They need healthy leaves and stems to grow too.

✏️ A tree has a very thick stem. What is it called?

Trees have the same parts as smaller plants. To stay healthy they need to have water and light like other plants. Trees have very long roots that can go deep in the ground to find water. They can grow very tall so they are above other trees to find light.

Some trees can store water in their trunks so in times of drought they can use the water to keep them healthy.

Amazing fact

The baobab tree can grow very big and very old. They can live for 3000 years and they can hold thousands of litres of water in their trunks. Sometimes the baobab tree is called the upside down tree because the branches look like roots.

📏 The diagram shows a tree and its roots. Measure the height of the tree with a ruler and write it in the box. Then measure the width of the roots and write it in the box. What do you notice?

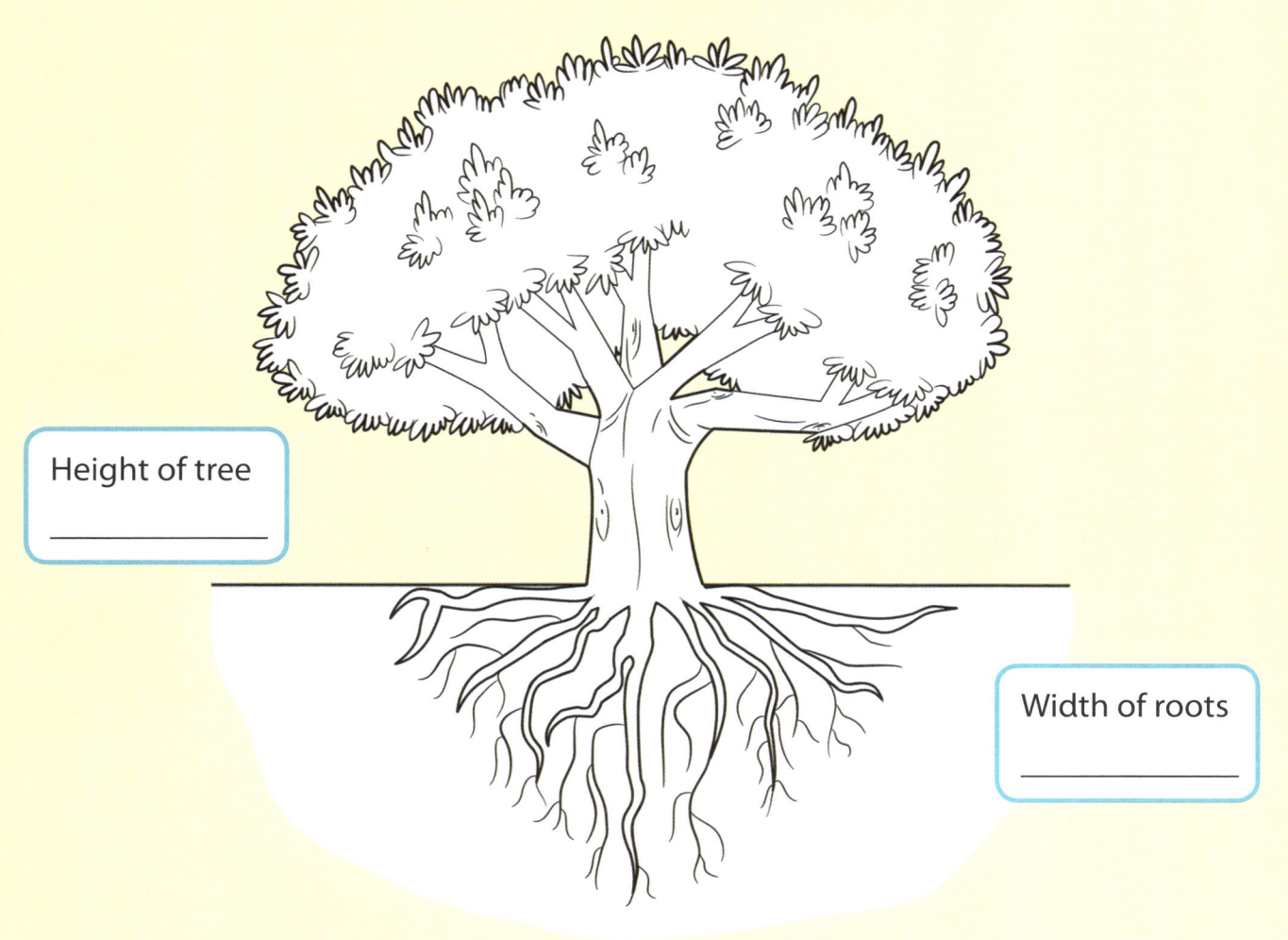

Height of tree
_____

Width of roots
_____

✏️ Why do you think trees have such big roots?

✏️ Circle the words that are needed for plants to grow well.

(healthy roots)   light   dark

unhealthy stems   water   sleep

healthy leaves   Sun   ears

Flowering Plants

67

Now turn to page 71 to review and reflect on what you have learned.

Not too hot and not too cold!

Learn that temperature changes the way plants grow.

The Big Idea

Plants grow best when the temperature is just right for them.

💬 Look at the pictures of these plants and where they are growing. What do you notice? Do they look healthy?

Flowering plants grow best when the temperature is just right for them to grow. Some plants like to grow in hot temperatures and some in very cold temperatures. Most plants like warm temperatures. A plant that grows in hot temperatures cannot survive in very cold temperatures. A plant that grows in cold temperatures cannot survive in very hot temperatures.

✏️ What has happened to this plant?

✏️ Is it too cold or too hot for this plant?

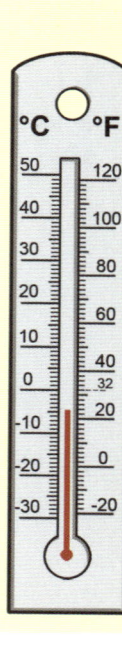

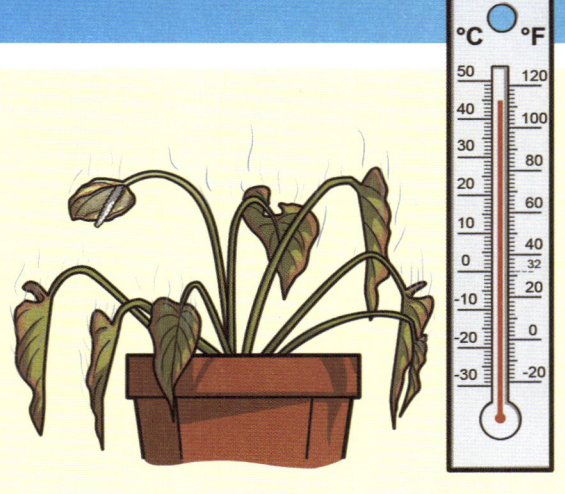

✏️ What has happened to this plant?

✏️ Is it too cold or too hot for this plant?

Greenhouses and polytunnels are used to grow vegetables, fruits and flowers. The air inside a greenhouse and polytunnel warms up quickly because hot air is trapped inside the glass or plastic. This means that plants which like warm weather can grow in the colder months.

💬 Why is this helpful to vegetable, fruit and flower growers?

Think about how you would carry out an investigation to test how temperature affects plant growth.

✏️ How would you make it a fair test?

Tick the things you would do in a fair test.

Cross the things you would not do.

1 Make sure the plants are the same size. ✓

2 Give the plants the same amount of water. ☐

3 Give the plants different amounts of light. ☐

4 Make one experiment last longer than the other. ☐

Look at the pictures and decide how you would help the plants to survive.

1

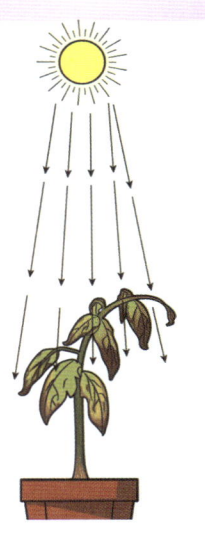

2

Flowering Plants

69

Now turn to page 71 to review and reflect on what you have learned.

What we have learned about flowering plants

Parts of a flowering plant

✏️ The flower is the name of one of the four parts of a flowering plant. Can you name the other three?

✏️ Describe a flower you have seen. What did it look like? What did it smell like?

I can name the four parts of a flowering plant ○

I can identify the different parts of a flowering plant. ○

What do plants need so they can grow?

✏️ Name two things that a plant needs to grow.

✏️ Two stages of an investigation are 'Asking questions' and 'Predicting what will happen'. Can you name two more stages?

✏️ What do plants take from the air?

I know what flowering plants need for healthy growth. ○

I know the stages of a science investigation. ○

70

How do plants take in water?

✏️ What do roots have that help them take in water from the ground?

✏️ Describe the journey that water takes after it has entered the roots.

I understand how water from the ground reaches the leaves. ○

Healthy plants

✏️ Name one thing that tells you that a plant is unhealthy.

✏️ Name two things that tree roots do for a tree.

I know when a plant is unhealthy. ○

Not too hot and not too cold!

✏️ What happens to a plant if the weather is too hot for it?

✏️ What happens to a plant if the weather is too cold for it?

I understand why healthy plants need to be not too hot and not too cold. ○

71

4 Introducing Forces

In this module you will:

- learn that pushes and pulls are examples of forces
- learn that we can measure forces with forcemeters
- describe how forces can change the shape of objects
- explore how forces can make objects start or stop moving
- understand how forces can make objects move faster or slower or change direction.

Amazing fact

The Great Pyramid in Egypt has 2 300 000 blocks of stone.
Each of the blocks weighs 2.5 tonnes. That is the same as two cars!
Some of the larger blocks weigh 50 tonnes.
In total 4000 people pulled and pushed 6 000 000 tonnes of rocks to build the Great Pyramid. It took about 30 years.

Physics

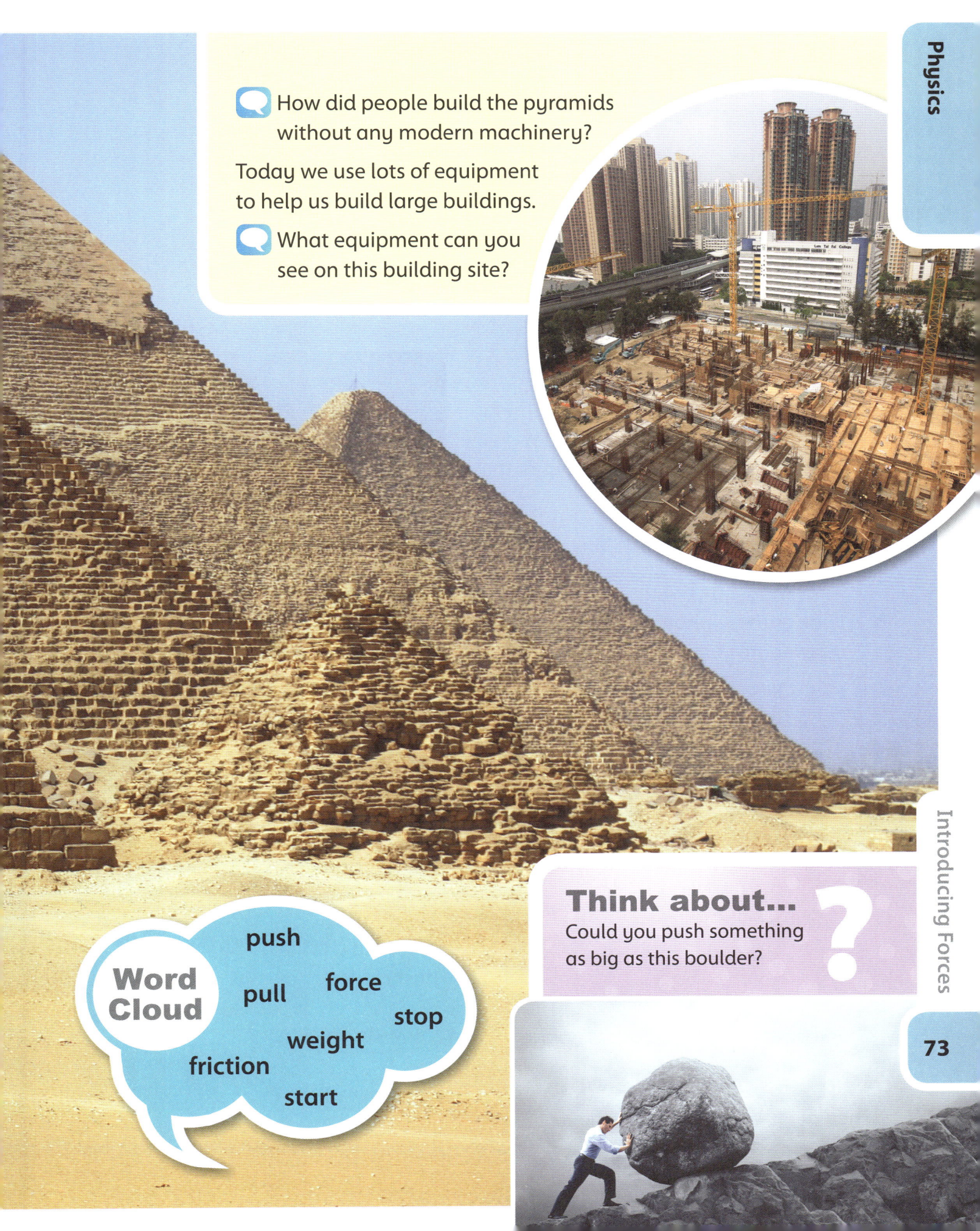

💬 How did people build the pyramids without any modern machinery?

Today we use lots of equipment to help us build large buildings.

💬 What equipment can you see on this building site?

Word Cloud
push
pull
force
stop
weight
friction
start

Think about...
Could you push something as big as this boulder?

Introducing Forces

73

Pushes and pulls

Learn that pushes and pulls are examples of forces.

The Big Idea

Pushes and pulls are forces.

💬 Have you pushed or pulled something today? Think about when you got dressed and when you went out of the bedroom.

A **force** is a **push** or a **pull**. You cannot see the force but you can see what it does.

✏️ Look at these photographs and answer the questions.

1 a What are the small boats doing? Circle the correct word.

They are pulling / pushing the crane.

b How do you know?

2 a What is this child doing? Circle the correct word.

The child is pulling / pushing the pushchair.

b How do you know?

74

We can use arrows to show the direction in which a force is acting. If an object is moving to the right, then the force must be acting to the right and we can draw an arrow to the right. If it is moving to the left, we can draw an arrow to the left. A force can go in any direction.

This is a drawing of the small boats and the crane. Draw an arrow on the drawing to show the direction of the force acting on the crane.

This is a drawing of the child and the pushchair. Draw an arrow on the drawing to show the direction of the force acting on the pushchair.

Do you think there are other forces acting on the crane and the pushchair? What is keeping the crane afloat? What is stopping the pushchair from floating away?

Some push and pull forces are very small and some are very big.

This butterfly is using its wings to push the air so it can fly. The force is so small we would not be able to feel it.

The photo below shows a force we would definitely be able to feel. The road has been damaged by a big earthquake. The push and pull forces of the earthquake were very strong.

Think about...
How can different forces be measured?

Pushes and pulls

Learn that we can measure forces with forcemeters.

The Big Idea

How can we measure a force if we cannot see it?

💬 **What force do we use to open a door?**

We open doors all the time. There are doors to get into our home, the classroom and cupboards and lockers. How much force is needed to open a door and how can we measure it? We can use a forcemeter.

💬 **Look closely at the picture of the forcemeter. Can you see why a forcemeter is also called a spring balance?**

✋ **Investigation: How does a forcemeter work?**

1. Your teacher will give you objects of different sizes.
2. Attach each object to the forcemeter.
3. Notice where the pointer reaches on the forcemeter.
4. Put the objects in order of how far they made the pointer go down.

✏️ **What did you notice?**

Weight is a force and it is measured in newtons (N) The bigger the mass of an object the greater its weight will be. Mass is measured in kilograms (kg) and grams (g).

Look at the photograph of the forcemeter and find 'NEWTONS' and the scale. Some forcemeters also show the mass in kilograms or grams.

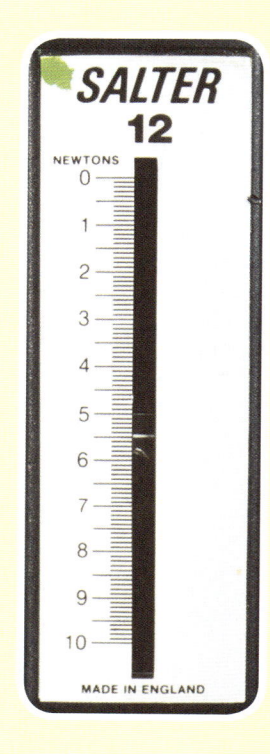

✏️ Work out the force in newtons and the mass in grams shown on the two scales.

76

✋ Investigation: How much force is needed to open a door?

1. Attach the hook on the forcemeter to the door handle of a closed door.
2. Hold the other end of the forcemeter and pull.
3. Look at the forcemeter to see how far the needle goes down. Read the amount if you can.
4. How much pulling force did you need to open the door?
5. Investigate other doors.

Now copy and complete the table in your Investigation notebook.

Location of the door	Force (N)

💬 What did you find out from your investigation? What force is needed to open a door?

An astronaut on the Moon has to wear heavy boots to make sure he does not float away. The gravitational force of the Moon is weaker than the Earth's so he weighs less there, even though his mass is the same.

Amazing fact

Gravity is the force which makes things have weight. The bigger something is, the greater its gravitational force. The Earth is very big so it has a very strong gravitational force. Gravity always pulls towards the centre of the Earth. That is why, if you throw a ball in the air, it always comes back down.

✏️ 1. What piece of equipment can you use to measure a force?

2. What unit do we use to measure a force?

3. How can you show the direction of a force?

Now turn to page 88 to review and reflect on what you have learned.

Making shapes with forces

Describe how forces can change the shape of objects.

The Big Idea

Forces can change the shape of an object.

💬 Crumple a piece of paper into the smallest ball that you can. What force have you used to do this?

Forces change the shape of objects. This can be very useful. Look at what the people are doing in the photos. They are using push and pull to make objects we can use and eat!

✏️ What are the people using to change the shape of the materials?

✋ **Investigation: Using forces to change the shape of modelling clay**

1. Use a wooden or plastic hammer and gently see what shape you can make out of a lump of clay.
2. Use your hands to make a pot out of a lump of clay.
3. Use your hands to push and pull the clay to make a flat pizza shape.
4. Draw 'before' and 'after' pictures of each of your models in your Investigation notebook.

The force of push is used to make parts for cars and planes. Machines press metal into moulds. They use the pushing force to create shapes from the flat metal.

This car body has been made in a mould.

✋ **Investigation: Using forces to mould modelling clay**

1. Press modelling clay into different moulds.
2. Take the clay out of the moulds. Did the moulds make good shapes?

Can you work out how these fruit tarts were made?

Describe the push and pull forces that were used.

Now turn to page 88 to review and reflect on what you have learned.

Forces can stop things moving

Explore how forces can make objects start or stop moving.

The Big Idea

When an object **starts** or **stops** moving, forces are used.

When is it useful to make an object stop?

Forces make things go faster and slow down or stop.

✋ **Investigation: Using forces to make a toy car change speed or stop**

Use a toy car to investigate how forces make the car change speed and stop.

1. Gently push a toy car along a smooth, flat surface such as the classroom floor.
2. Observe how the car moves.
3. Measure how far the car travels before it stops.

✏️ Record the distance in the space below.

My car travelled _____ cm.

✏️ Did the car go quickly or slowly?

_____

✋ **Investigation: Changing the speed and direction of a toy car**

1. Now push the car as hard as you can along the classroom floor. Make sure nobody is in front of the car.
2. Observe the car travelling.
3. Measure how far the car travelled.

✏️ Record the distance in the space below.

Did the car change direction?

My car travelled _____ cm.

It _____ direction.

✏️ Did the car go more quickly or slowly this time?

_____

✏️ Tick the sentence that is correct.

The more force there is, the faster the car goes. ☐

The less force there is, the faster the car goes. ☐

Introducing Forces

81

Forces can stop things moving

Explore how forces can make objects start or stop moving.

The Big Idea

Different surfaces affect how fast or slowly objects move.

💬 **What makes things go faster?**

How fast things go and how quickly they stop also depends on the surface they are travelling along.

💬 **How fast do you think the vehicle in this photo can go? What is stopping it from going fast?**

✋ **Investigation: How well do things move on different surfaces?**

We can investigate how things travel along different surfaces using a ramp and different materials.

1. Make a ramp out of cardboard, guttering or a piece of wood.
2. Put some books or a box on the floor and lean the ramp against them.

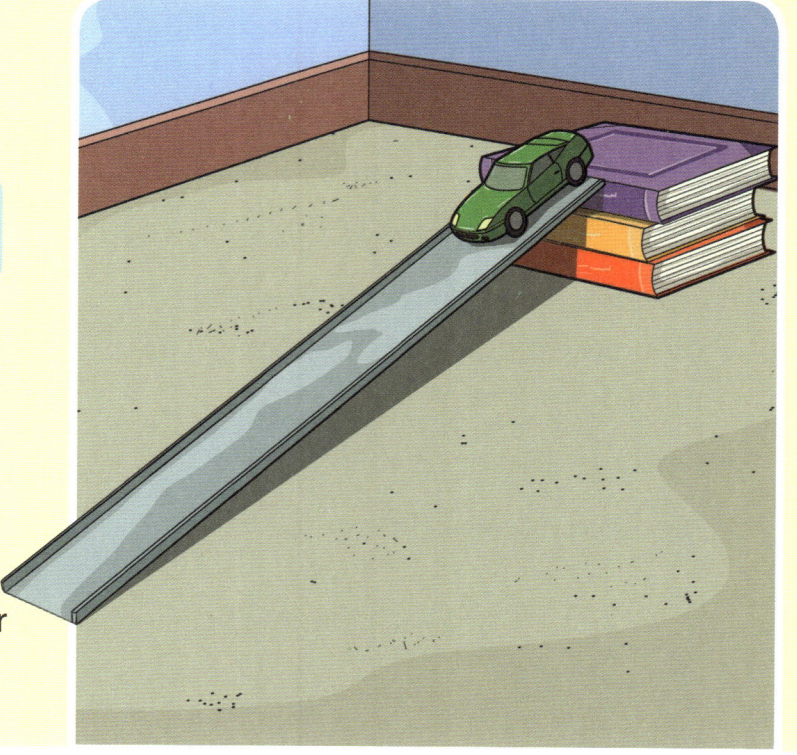

3 Hold your car at the top of the ramp and let go.

4 Put the ramp on surfaces made of different materials, such as soil, concrete and carpet. Measure how far the toy car travels on each surface.

5 Copy and complete the table in your Investigation notebook

Type of surface	Distance travelled (cm)	Observations

Which surface made the car travel the fastest?

Which surface made the car travel the furthest?

Think about...
Why does the rough surface make the car move more slowly?

Do all objects move in the same way as the toy car?

Try rolling a ball down the ramp. Does the ball travel faster than the car or more slowly? Do the different surfaces make the ball behave in the same way as the car?

Do the objects go faster on a smooth or a rough surface?

What happens to a toy car if you do not push it?

What happens to the toy car when you push it very hard?

What happens when the car goes across a rough surface?

Now turn to page 89 to review and reflect on what you have learned.

Forces can affect speed and direction

Understand how forces can make objects move faster or slower or change direction.

The Big Idea

We wear different types of shoes for different purposes.

💬 Think about what we have learned about surfaces. What kind of surface is best for a children's slide?

Some surfaces make objects travel quickly. This is very useful if you are making a slide or you want to go ice skating.

Some surfaces make objects travel more slowly. This is useful if you want cars to go more slowly, as we found out on page 83.

✋ Investigation: What slows things down?

1. Put your hands palm down on the surface of your desk and press down.
2. Move your hand across the desk. What do you feel?

💬 Did you feel a dragging feeling?

Friction

This dragging feeling is called **friction**. Friction always happens when two surfaces rub together. Look again at the photo of the ice skaters above and the vehicle on a rough surface on page 82. Which surface do you think has the most friction?

When two surfaces rub together friction gives us grip. Some shoes have good grip to stop us falling on slippery ground. Walkers have good grip on their boots to stop them slipping.

💬 Who else wears shoes with a good grip?

✏️ Look at the pictures of different shoes. Circle the ones you think have the best grip.

💬 How could you investigate this? Think of an investigation you could do to test which shoe has the best grip.

✏️ Write or draw a picture explaining what you would do in your Investigation notebook.

Brakes

💬 Do you have brakes on your bike? What do they do?

Brakes use friction to stop moving things quickly. When you press the brake lever on the bike, the brake blocks squeeze against the bike wheel, stopping the bike wheel from moving.

Forces can affect speed and direction

Understand how forces can make objects move faster or slower or change direction.

The Big Idea

Why do objects change direction?

✏️ When you throw a ball at a wall, what force makes it go forward?

✏️ What do you think will happen to the ball when it hits the wall?

✋ Investigation: What happens when a moving object hits another object?

1. Sit on the floor in front of a wall or door. Carefully roll the ball towards the wall. What happens?
2. With a partner, both roll a ball towards each other so the balls hit. What happens?
3. Try this with toy cars. What happens?
4. Write what you found out in the table below.

Experiment	What happened?
Ball rolled to a wall	
Two balls rolled towards each other	
Two toy cars pushed towards each other	

💬 What do you think is happening when an object hits another object?

✏️ Complete the sentences using the words in the word bank.

When two moving objects hit _____ the opposite forces can make the objects change _____.

When two objects hit each other _____ slows the objects _____.

Word Bank

**direction   friction   down
each other**

A footballer pushes the ball when he kicks it. Footballers can use forces to decide where they want to kick a ball. They make decisions about how fast the ball is coming towards them, what angle it is and how much spin it has.

✏️ Answer the questions about forces.

1 Name the force that gives us grip on our shoes.

2 Give an example of a force that changes the direction of an object.

3 When is friction useful?

Introducing Forces

87

Now turn to page 89 to review and reflect on what you have learned.

What we have learned about introducing forces

Pushes and pulls

✏️ What does N stand for?

✏️ What units do we use to measure mass?

✏️ Is gravity stronger on Earth or the Moon?

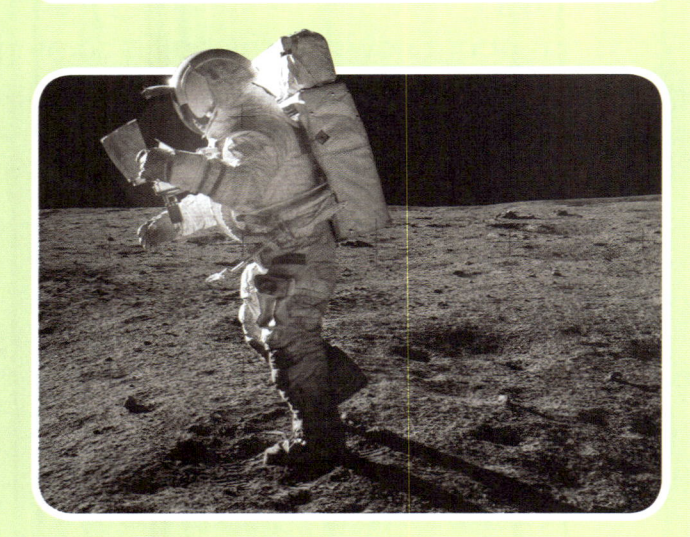

I know that pushes and pulls are forces. ◯

I know that forces are measured in newtons. ◯

Making shapes with forces

✏️ Name two kinds of force that change the shape of an object.

✏️ What is a mould?

I understand that different kinds of pushing and pulling forces can change the shape of an object. ◯

88

Forces can stop things moving

✏️ What happens to a toy car when you increase the pushing force?

✏️ How can you change the direction of the toy car?

✏️ How can you make something stop?

I understand that a pulling or pushing force makes things move. ○

I understand that a pulling or pushing force makes things stop. ○

Forces can affect speed and direction

✏️ What is the name of the force between two surfaces that slows things down?

✏️ Can you name somewhere where we use this force to help us?

✏️ What happens when two moving objects hit each other?

I know that friction is a slowing down force. ○

I understand that the greater the force, the faster things move. ○

I understand that changing the direction of the force changes the direction of the movement. ○

Introducing Forces

89

5 The Senses

In this module you will:
- explore our senses and how we use them to learn about the world around us
- learn about touch
- learn about taste
- learn about smell
- learn about sight
- learn about hearing.

Word Cloud: taste, smell, hear, sight, touch, sense

I can see

These are our senses. We have them to protect ourselves and to find out about what is around us. We have looked at how animals use their senses. Now we are going to explore how we use our senses of touch, taste, smell, sight and hearing.

- Which senses are the children using to make sure they cross the road safely?
- How does the sense of touch protect us?

Imagine you have no sense of touch. What will happen?

I can **smell**

I can **taste**

I can **hear**

I can **touch**

Biology

The Senses

91

Amazing fact

Some animals have a sense of smell that is hundreds of times better than a human's sense of smell.

Touch

Explore our senses and how we use them to learn about the world around us. Learn about touch.

The Big Idea

Touch is a **sense**.

✏️ If you want to touch something, which part of your body do you use?

You feel with your fingers, but the skin on all parts of the body can feel too. When you are hot or cold your whole body feels this.

💬 Senses protect us. How does the sense of touch protect us? Look at the picture. What is happening?

The sense of touch warns the person to move their hand away quickly before they get badly burned.

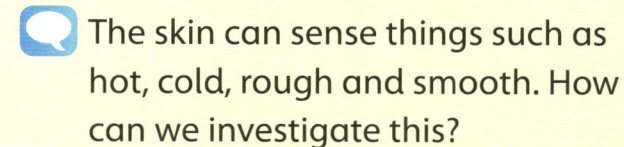

💬 The skin can sense things such as hot, cold, rough and smooth. How can we investigate this?

92

✋ **Investigation: Feeling objects**

1 Work in pairs. Blindfold your partner or ask them to close their eyes.

2 Direct your partner's hand to different objects on the desk.

3 Explore how well your partner can identify if the objects are rough, smooth, warm or cold.

4 Sort the objects and then copy and complete the table in your Investigation notebook.

Warm things	Cold things	Rough things	Smooth things

How easy is it to use touch to work out who a person is? Try it out.

✋ **Investigation: Blind man's buff**

1 Ask one person in your group to volunteer. Blindfold them – they are now 'it'.

2 Everyone else stands quietly without moving.

3 The blindfolded student walks to a person in the group. They use their hands to work out who the person is.

4 If the blindfolded student gets it right, the person they identified becomes 'it'.

Was it easy?

Amazing fact

The thickest skin on your body is on the palms of your hands and the soles of your feet. It can be about 1.5 mm thick.

Think about...

How important is the sense of touch to people who have no sight or poor sight?

The Senses

93

Touch

Explore our senses and how we use them to learn about the world around us.

Learn about touch.

The Big Idea

You can find out a lot of information using touch.

Simple loops

Double loops

Whorl

arch

💬 Look closely at the ends of your fingers. What do you see?

At the ends of your fingers there is a pattern. Look at the picture. Do any of the patterns on your fingers match these?

To make it easier to see the pattern, we are going to carry out an investigation. We are also going to find out whether everyone's patterns are the same or different.

✋ Investigation: Fingerprint patterns – are they all different?

1 Press your finger on the ink or paint. Do not get too much paint or ink on your finger. If you use just the right amount of paint you will be able to see the pattern of your print clearly.

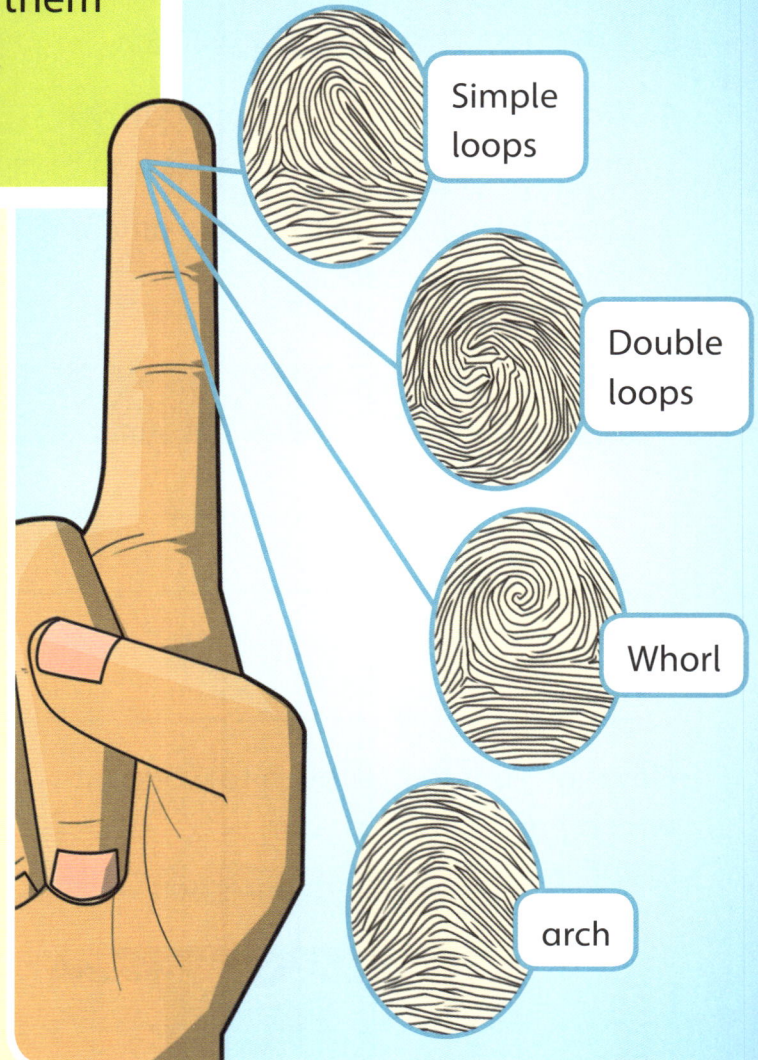

2 Press your finger on to a piece of paper.

3 Make a print with all your fingers and your thumb. Leave your fingerprints to dry.

✏️ Which of the patterns can you see?

✏️ Look at everyone's fingerprints and complete the table. Use tally marks.

Pattern	Number of students
Arch	
Whorl	
Double loop	
Simple loop	

✏️ **1** Which pattern did the most students have?

2 Which pattern did the fewest students have?

3 What did you find out? Circle the correct word in the sentence.

Everyone's fingerprints looked the same / similar / different.

Because everyone's fingerprint is different, we can use fingerprints to identify a person.

The girl in the picture is using her fingerprint to identify who she is. This is how she pays for her school lunch.

This detective is matching up fingerprints to see if he can find out who a burglar is.

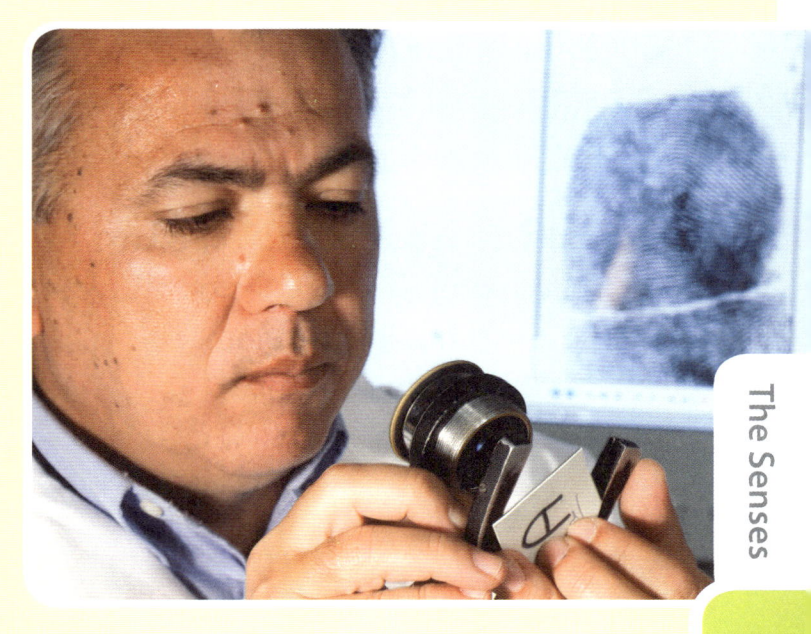

Think about...
Why do we have the sense of touch?

The Senses

95

Touch

Explore our senses and how we use them to learn about the world around us.

Learn about touch.

The Big Idea

Touch helps to protect us but can sometimes confuse us.

💬 Remember the person who burned their finger on page 92. What did their sense of touch tell them?

We use our sense of touch to identify things and to protect us. But sometimes our sense of touch can trick us.

✋ Investigation: Does the sense of touch sometimes trick us?

1. Work with a partner. You need a container of warm water, a container of room temperature water and a container of cold water.

warm water

room temperature water

cold water

2. Place the containers in the order shown.

3. Put your right hand in the cold water while your partner puts their left hand in the warm water. Leave your hand in the water for 1 minute. Then both put your hand in the room temperature water.

✏️ Complete the sentences in the speech bubbles. Write what you and your partner felt. Use the word 'cooler' or 'warmer'.

Me

When I put my hand in the room temperature water it felt _____.

My partner

When I put my hand in the room temperature water it felt _____.

💬 Did the room temperature water feel the same for both of you or did it feel different?

Swap roles and try the investigation again. What happened this time? Talk about what is happening.

How good are our fingers at touching when we cannot see? We are going to carry out an investigation to find out.

Investigation: Sorting sandpaper

1 Work in pairs. Sit at your desk and your partner will blindfold you.

2 Your partner will place different pieces of sandpaper in front of you. They have been numbered in order of roughness. Can you sort them into the correct order using just touch?

3 Take off your blindfold to see if you have put them in the correct order. How did you do?

4 Now let your partner have a turn.

Write your order here:

roughest

☐ ☐ ☐ ☐ ☐

smoothest

Did you put all the sandpapers in the correct order?

Where did your fingers get confused?

Compare your order with your partner's.

Did your partner make the same mistakes as you?

Amazing fact

Sandpaper comes in different grades of roughness because it is used to rub surfaces down. Decorators use it to sand down paintwork. They start with rough sandpaper and finish with fine sandpaper to make a smoother surface.

Complete the sentences using the words in the word bank.

Our sense of ___touch___ protects us from _____. We use our _____ to feel things but every part of our skin can feel things. We all have different _____.

Word Bank

danger   fingers
fingerprints   ~~touch~~

The Senses

97

Now turn to page 114 to review and reflect on what you have learned.

Taste

Explore our senses and how we use them to learn about the world around us.

Learn about taste.

The Big Idea

We use our tongues to **taste** things.

💬 Do you like sweet things? Or do you like sour things?

What is your favourite taste?

We need to eat lots of different kinds of food to stay healthy. These foods have many different flavours. The tongue is covered in tiny taste buds that help us to taste.

All the foods that you can think of are in the four taste groups: sweet, bitter, sour and salty. Some foods have a mixture of different tastes. Some scientists include savoury food as a fifth group.

Find out which parts of your tongue have the most taste buds.

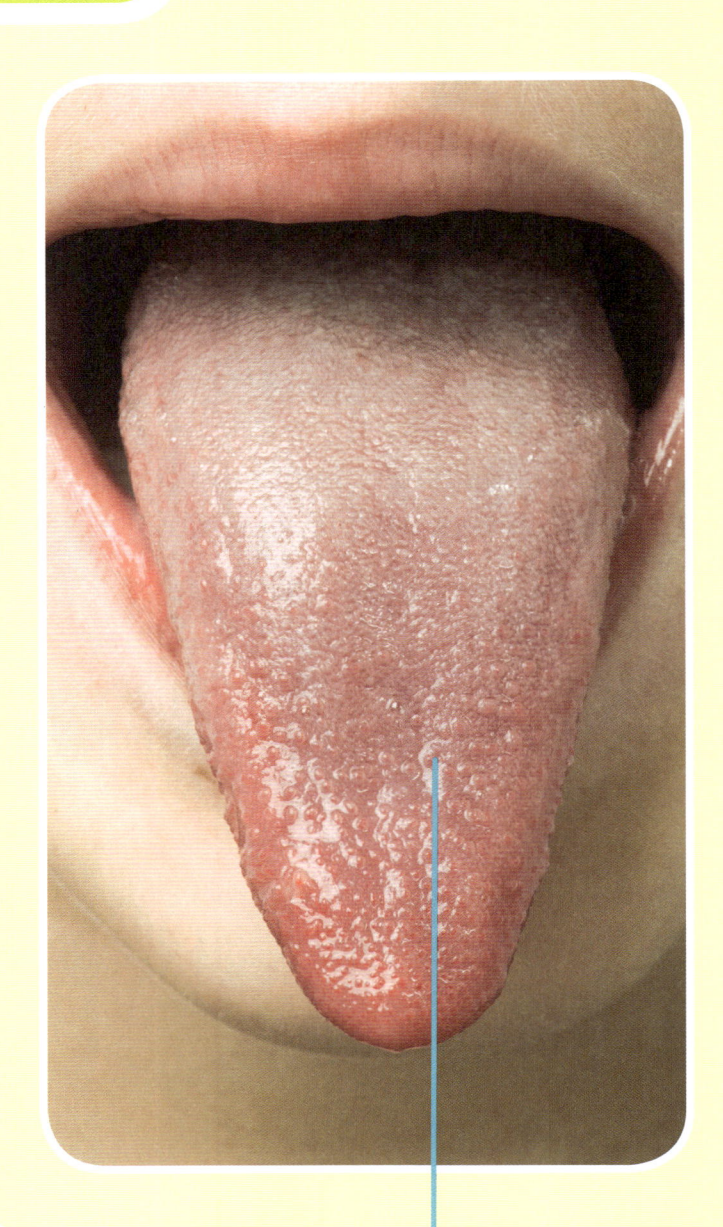

Taste buds

Amazing fact

We have about 10 000 taste buds in our mouth and, in general, girls have more than boys!

✋ **Investigation: Which parts of the tongue have the most taste buds?**

1 Put a lolly stick in the cup of flavoured water.

2 Place it on different parts of your tongue:
- the front
- the top
- the sides
- under your tongue.

3 Colour the tongue to show where you tasted the flavour most.

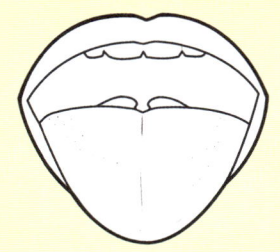

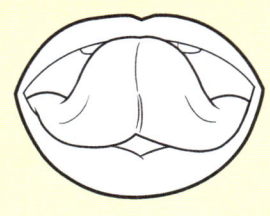

💬 Compare your tongue map with your group.

Does it look the same or does it look different?

✏️ Which parts of your tongue tasted the flavour the most?

⚠️ The back of the tongue also has lots of taste buds. We did not test this because it is not safe to put objects in the back of your mouth.

Some people are allergic to some foods, such as peanuts.

There are sweet, salty, sour, bitter and savoury foods. Which foods are most popular?

✏️ Carry out a survey of your class. Find out which is the most popular taste. Copy and complete the table using tally marks.

Taste	Number of students
Salty	
Sour	
Sweet	
Bitter	
Savoury	

✏️ Show the results as a bar chart, using the information you recorded in the table.

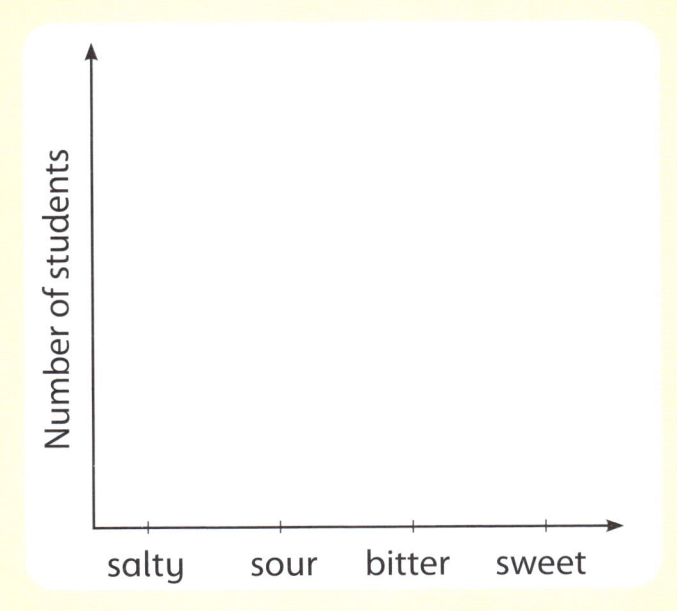

The Senses

99

Taste

Explore our senses and how we use them to learn about the world around us. Learn about taste.

The Big Idea

Taste tells us which foods we need to stay healthy.

✏️ Can you remember the five different taste groups? Unjumble the words.

w e t e s

y l t a s

r u o s

t r i b e t

o u r v a s y

We eat food because we like it. There are lots of different foods. Taste tells our body what the food will do for us.

Sweet foods give us energy. Sweet foods are called carbohydrates.

Savoury foods often contain protein. Protein helps us grow.

We need some salt in our bodies to keep us healthy.

Sour and bitter foods often contain vitamins, which keep us healthy.

100

Investigation: Sorting food into groups

Your teacher will give you some different foods. Work with a partner to sort the foods into the different food groups in the table.

Carbohydrates	Proteins	Salty food	Vitamins

Saliva

Saliva is the liquid in our mouths. It is useful because it helps break down the food we eat. We are going to test how saliva helps us to taste things.

Investigation: Does saliva help us taste?

1. Stick out your tongue to dry it.
2. Put a few grains of salt on your tongue. What can you taste?
3. Rinse your mouth. Dry your tongue again and place some sugar granules on your tongue. What can you taste?
4. This time do not dry your tongue. Wait a few minutes until you have lots of saliva. Now place the salt on your tongue. Rinse your mouth, then place the sugar on your tongue. What can you taste this time?

Complete the sentence.

With a dry tongue I could / could not taste the sugar and salt as well as I could with saliva.

How does taste protect us?

1. Where are our taste buds?

2. List two foods that help us grow.

3. Does saliva help us taste food?

Now turn to page 114 to review and reflect on what you have learned.

Smell

Explore our senses and how we use them to learn about the world around us.

Learn about smell.

The Big Idea

The sense of **smell** is useful to us.

💬 Close your eyes. What can you smell?

We use the sense of smell to find out if food is good or bad. We also use the sense of smell to make us feel good.

Some people wear perfume to make them smell nice. Bad smells make us feel unwell. We move away from bad smells because they might hurt us or make us ill.

💬 Look at the photos. Do you think they show something that smells nice or not? Did you all agree?

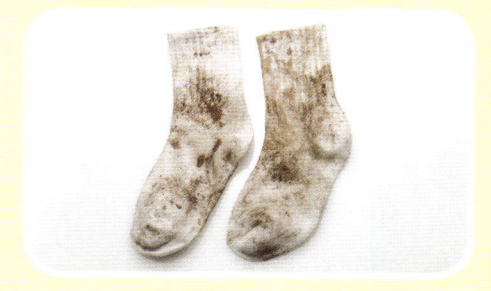

Amazing fact

Everyone has their own personal smell. It is like a giant fingerprint.

Think about...

What smells would you miss if you lost your sense of smell?

✎ What do we use to smell things?

The cells send a message to the brain, which works out what the smell is.

We breathe air through these holes called nostrils.

In the nostrils there are millions of very small cells called receptors that smell the smells in the air we breathe in.

How quickly do our brains recognise smells? Try this investigation.

✋ **Investigation: Recognising smells**

1 Make a smell box. Take a small clean container like a yogurt pot.
2 Choose something that has a strong smell. Place it in the container.
3 Cover the container with plastic or paper. Use tape to seal the lid in place. Make sure you cannot see what is in the container.
4 Carefully make holes in the lid.
5 Ask the students in your class if they can recognise the smell. How long did they take: quick, medium or slow?

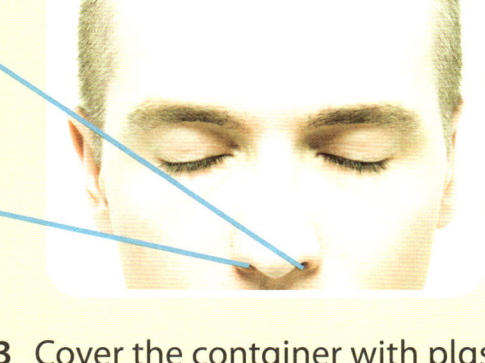

Copy and complete the table in your Investigation notebook.

Smelly object	Name of student	Did they recognise the smell correctly?	How long did it take them?

✎ How many students guessed the smell?

✎ How many students did not guess the smell?

The Senses

103

Smell

Explore our senses and how we use them to learn about the world around us.
Learn about smell.

The Big Idea

The sense of smell can save us from danger.

💬 How does the nose sense smell?

The nose gives us information about our world just like the senses of taste and touch. Smells can warn us of danger.

💬 What are the dangers in the photos of the building and the fruit? How can our sense of smell warn us?

We know that the smell of smoke can travel, but do other smells travel too? Have a go at finding out.

✋ Investigation: How far does smell travel?

1. Sit very still. Someone will spray something in the room.
2. Put your hand up when you can smell the spray.
3. Work with a partner to investigate how far the spray can be smelled.

✏️ What is the furthest distance that you could still smell the spray?

💬 Compare and discuss your results with the rest of the class.

Smell is important because it helps us to work out the taste of things. Find out how important the sense of smell is by carrying out an investigation.

✋ **Investigation: Do we use smell when we taste things?**

1. Hold your nose or put a peg on it.
2. Now eat a slice of orange.
3. Then eat a slice of orange without holding your nose.

💬 What did you notice? Compare what you found with your partner. Did they notice the same thing?

✏️ Is taste important or not important when we taste things? Write your answer in the box.

Amazing fact

The smell of smoke from some wildfires can travel thousands of kilometres.

⚠️ Be careful if you spray perfumes. Some people can be allergic to them.

✏️ 1 Match up the words with their meanings.

Nostrils	a nice smell
Receptors	part of the face which we use to smell
Nose	holes in the nose
Perfume	small cells in the nostrils

2 How does smell protects us? Write about two ways.

Now turn to page 115 to review and reflect on what you have learned.

Sight

Explore our senses and how we use them to learn about the world around us.

Learn about sight.

The Big Idea

Seeing things helps us.

💬 What was the first thing that you saw this morning?

Like all our senses, our **sight** tells us about the world around us.

Some animals have eyes at the side of the head, like this horse. This helps them see if there is any danger.

Amazing fact

About 95 per cent of animals have eyes. Why do some animals not have eyes?

Humans and many other animals have eyes at the front of their head. This helps us see things in front of us in a lot of detail.

💬 Do you need two eyes to see?

✋ Investigation: Is seeing with two eyes better than one? Part 1

1. Hold a pencil in each hand and stretch your arms out.
2. Close your left eye and try to make the ends of the two pencils meet.
3. Now close your right eye and try to make the ends of the pencils meet.

💬 What happens? Is it easy to make the pencils meet?

4. Now try this again with both your eyes open.

💬 Does this make a difference?

✋ **Investigation: Is seeing with two eyes better than one? Part 2**

1. Sit your partner at a desk with a paper cup on the desk about an arm's length away from them.
2. Ask your partner to close their left eye.
3. Hold a paperclip in your hand. Move the paperclip slowly over the cup about an arm's length above it.
4. Your partner shouts 'drop it' when they think the paperclip will fall directly into the cup.
5. Repeat with both eyes open.

Swap roles and repeat the investigation.

💬 Does using two eyes make it easier to get the paperclip into the cup? Did it make it easier for everyone?

✋ **Investigation: Reading with one eye closed**

1. Close your left eye and point to a word.
2. Open your left eye and close your right eye.
3. Now look at the word with both eyes.

💬 What happens to the position of your finger?

✏️ What have you discovered in your investigations? Is it better to have two eyes or one?

The Senses

107

Sight

Explore our senses and how we use them to learn about the world around us.

Learn about sight.

The Big Idea

You do not always see what you think you see.

💬 How far can you see? How well can you see in the dark?

Our sight is not as good as lots of other animals. To be able to see we need light. The light from the object we are looking at hits the lining at the back of our eye which has special cells. The cells send a message to our brain and the brain works out what we are looking at. One part of this lining does not have any cells and is called the blind spot. If the light from an object hits this part of the eye, we cannot see it.

💬 How can we show that we have a blind spot?

✋ Investigation: The blind spot

•   x

1. Hold the card in front of you. Close your right eye.
2. Stare at the X with your left eye. Slowly move the card closer to your face.
3. Keep looking at the X with your left eye. The dot will disappear. This is your blind spot.
4. Repeat this test with the other eye. This time close your left eye. Stare at the dot with your right eye. Slowly move the card closer until the X disappears. You have found your blind spot.

Amazing fact

Hawks and eagles have very good sight. It can be eight times better than a human's. A hawk can see a rabbit 1.5 kilometres away.

Our eyes work very closely with the brain. Sometimes the brain plays tricks on us. It does this to try to help us by filling in the blanks for us.

✋ **Investigation: Brain tricks**

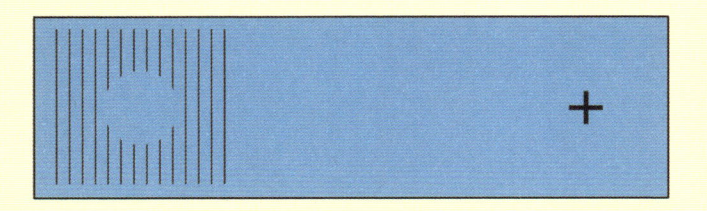

1. Hold the card in front of you.
2. Close your left eye.
3. Stare at the cross and move the card closer to your face.

✏️ What happens to the space in the middle of the lines?

💬 What happened to the space for everyone else?

Our brain recognises that something is missing. It thinks it is the blind spot and fills the space in for us.

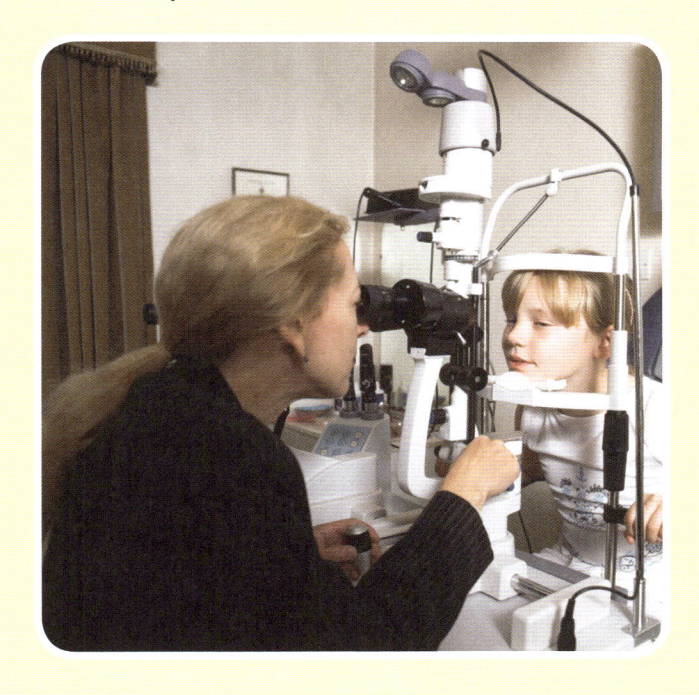

An optician checks that our eyes are working properly. If they are not, he can work out what kind of glasses we need. Have you ever had your eyes tested?

1. What do our eyes need to see properly?

2. Which has the better eyesight? Circle the correct one.

    Hawk   Human

3. Which is better – having two eyes or one? Circle the correct answer.

    One   Two

Now turn to page 115 to review and reflect on what you have learned.

The Senses

109

Hearing

Explore our senses and how we use them to learn about the world around us.

Learn about hearing.

The Big Idea

Are two ears better than one?

💬 Listen. How many different sounds can you **hear**?

The world is full of sounds. Sounds can warn us of danger. Sounds can make us happy. Sounds can make us sad.

We hear sounds through our ears. The vibrations made by an instrument or voice travel through the air into our ears. Then our brain works out what the sound is.

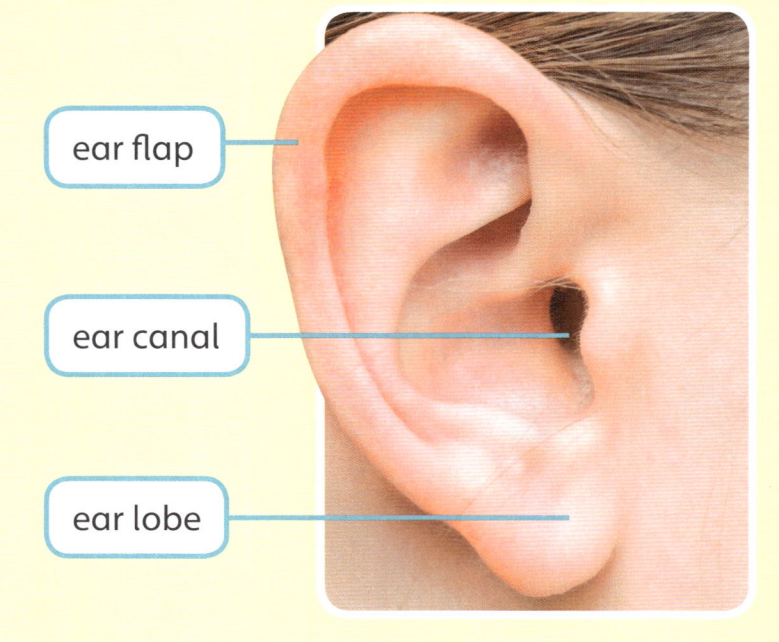

- ear flap
- ear canal
- ear lobe

Recognising different sounds should be easy. Try this investigation.

✋ Investigation: Guessing sounds

1. Collect some objects that make a sound.
2. Ask your partner to stand with their back to you.
3. Pick up one of the objects and make a sound with it.
4. Record whether your partner recognised all the sounds. Copy and complete the table in your Investigation notebook.

Object	Yes	No

Think about...
Why do some animals have very large ears?

Now swap places. Compare your results with your partner.

💬 Did you get all the sounds correct? Did you both get the same ones wrong?

💬 How can we find out if two ears are better than one?

✋ Investigation: Direction of sounds

1. One person sits on the floor with their eyes tightly closed.
2. The rest of the class sit in a circle around them.
3. One student in the circle says 'ears'. The person in the centre has to point to where the voice came from and say who said it.
4. They try again. First they cover one ear. Then they cup their hands behind both ears.

💬 Was it easier with one or two ears? Did it help when the student cupped their hands behind their ears?

Hearing

Explore our senses and how we use them to learn about the world around us. Learn about hearing.

The Big Idea

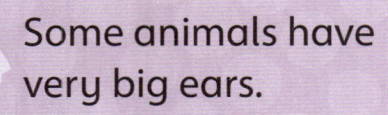

Some animals have very big ears.

✏️ Why do these animals have big ears?

In one of the investigations on page 111 you cupped your hands behind your ears to see if it helped you to hear better. This was like making your ears bigger. We can investigate ear size and whether it helps hearing.

✋ Investigation: Hearing and ear size

1. In groups make a funnel with a piece of card or paper.
2. Make more funnels in different sizes.
3. Choose something that makes a sound. Use this sound for all your tests.
4. Stand at the same distance from the sound each time.
5. Test all the funnels by putting them against your ear. Listen carefully with your eyes shut.

Remember
Do you remember what a fair test is? These are the things you should do in this investigation to make it fair:
- use the same sound
- stand the same distance away.

💬 Which funnel worked the best?

Morse code is a way of sending messages using different lengths of sound. Dashes and dots show whether the sound should be long or short.

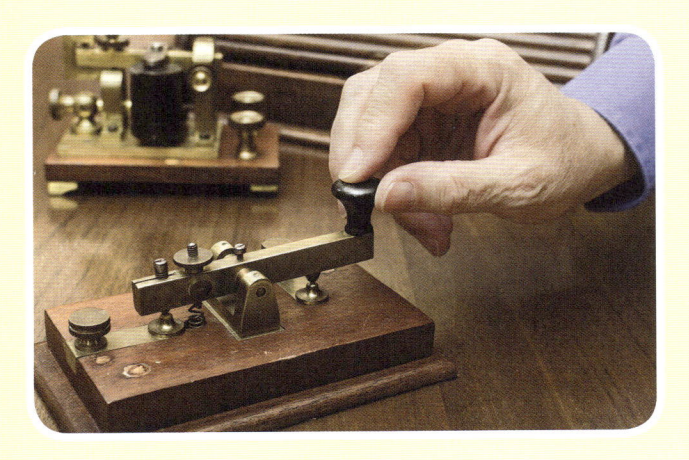

Letter of the alphabet	Morse code dots and dashes
A	. −
E	.
I	. .
O	− − −
C	− . − .
N	− .
M	− −
T	−
H	. . . .
S	. . .

✋ **Investigation: Investigating sound codes**

1. Work with a partner.
2. The table tells you how to make a letter using only sound. Use the palm of your hand for dash and your knuckles or fingertips for a dot.
3. Tap out the code for:

   C − . − .

   A . −

   T −

4. Take it in turns to send more messages to each other by making the sounds on your desk.

✏️ Complete the sentences using the words in the word bank.

We use our ___ears___ to hear things. We found out that using a _____ helped us hear _____. Elephants have big ears to keep them _____. Hares have big ears to trap the _____.

Word Bank

~~ears~~   sound   hearing trumpet   better   cool

The Senses

113

Now turn to page 115 to review and reflect on what you have learned.

What we have learned about the senses

Touch

✎ Which part of the body senses touch?

✎ Imagine you cannot see. Which senses will you use to find your way around?

✎ Our sense of touch protects us. Give an example of when the sense of touch can stop us from being hurt.

I know that the sense of touch tells me how things feel. ○

I know that the sense of touch can protect me from danger. ○

Taste

✎ How can the sense of taste protect us?

✎ Humans can detect five different tastes. Can you fill in the tastes that are missing?

sweet   bitter   savoury

I know that the sense of taste works better with the help of the sense of smell. ○

I know that the sense of taste can protect me from danger. ○

Smell

✏️ Where are your smell receptors?

✏️ Why do animals and some people smell their food before they eat it?

I know that the sense of smell can protect me from danger. ◯

Sight

✏️ Why do animals and people have two eyes at the front of the head?

✏️ What is the blind spot in our eyes?

I know that having two eyes is better than having one. ◯

I know that the sense of sight can protect me from danger. ◯

Hearing

✏️ Why does having large ears help some animals to hear?

✏️ Name two parts of the ear.

I know that having two ears is better than having one. ◯

I know that the sense of hearing can protect me from danger. ◯

The Senses

115

6 Keeping Healthy

In this module you will:

- revise the life processes: eating, drinking, moving, reproducing and growing
- understand how a varied diet and exercise keep us healthy
- learn that some food can harm our bodies.

Do you like running and playing games?

What foods do you like eating?

Word Cloud

human
nutrition
living
exercise
diet
healthy
food

✏️ Choose your favourite foods and drinks for breakfast.

I like to eat _____   I like to drink _____

I like to eat _____   I like to drink _____

I like to eat _____   I like to drink _____

I like to eat _____

Think about...
Why is food so important to us?

Amazing fact
In the world today there are approximately 75 women and 40 men who are over 110 years of age! What has helped them reach such an old age?

The life processes

Revise the life processes: eating, drinking, moving, reproducing and growing.

The Big Idea

We all eat, drink, grow and move.

What can you remember about the life processes we discussed in Module 1?

Which life processes do the photographs show? Use the words in the word bank to label the photographs.

Word Bank

growing   moving   eating   drinking   reproducing

118

All animals and **humans** have to eat food and drink water to stay alive and not only because they taste good! Eating and drinking to stay alive is called **nutrition**. If we do not eat and drink the life processes stop.

All animals and humans move to find food and drink. Animals move to get away from predators.

Animals and humans reproduce and have young so that their species survives.

Not everyone has enough food and water

In some parts of the world people do not have enough food and water to survive. Bad weather or natural disasters can destroy food crops. If a country is at war, it might not have enough money or enough people to grow crops.

Nobody wants to see people die of hunger and thirst so other countries give food and water.

✏️ Draw a poster in your Investigation notebook. Explain why children who are dying from hunger and thirst need food and water. Use the words in the word bank to help you.

Word Bank

| food | nutrition | healthy |
| water | help | give |

✏️ Write the missing letters to complete the words. They are the life processes.

m __ __ i n g   d __ i n __ i __ g

__ a t __ n __   g r __ __ n __

r __ p r __ d __ c i __ g

Keeping Healthy

119

Now turn to page 134 to review and reflect on what you have learned.

Diet and exercise

Understand how a varied diet and exercise keep us healthy.

The Big Idea

Different foods contain different substances which help us live and grow.

Some animals need to eat only one type of **food**. A lion might only eat meat but it gets everything it needs from the meat. Humans need to eat lots of different types of food.

✏️ Look at this plate. Write down some of the different foods you can see.

Scientists have grouped the foods by what they do for our bodies. The picture below shows what groups the foods belong to. The four main food groups are carbohydrates, fats, proteins, and vitamins and minerals. If we eat some foods from each of these groups we have a balanced diet and we keep healthy.

Vitamins and minerals

Carbohydrates

Fats

Proteins

This is what the food groups do for us.

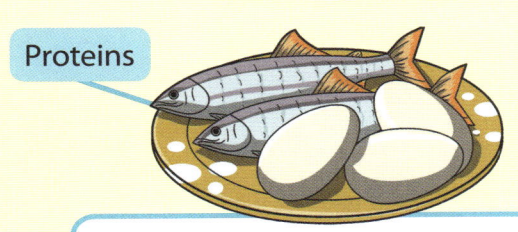

Proteins

We give you protein. Protein helps build muscles.

Carbohydrates

We give you carbohydrates. Carbohydrates give you energy.

Fats

We give you fat. Fat gives you energy and keeps you warm.

Vitamins and minerals

We give you vitamins and minerals. Vitamins and minerals keep your body healthy. You need minerals to build bone and teeth.

Amazing fact

Energy in food is measured in joules or kilocalories. A 9-year-old child needs about 1800 kilocalories a day to be healthy.

Think about...

What can happen if we do not eat enough of the right foods?

1. Imagine you feel unwell. Which food group might help you to feel better?

2. Imagine you are going to run a marathon. Which food group do you need to eat more of?

3. Imagine you want to build up your muscles. Which food group do you need to eat more of?

4. Imagine the weather is very cold. Which food group will help to keep you warm?

Keeping Healthy

121

Diet and exercise

Understand how a varied diet and exercise keep us healthy.

The Big Idea

If we do not eat the right foods we can become ill.

✏️ Can you remember the main food groups? Write them here.

1	2
3	4

Sometimes it is hard for humans to get the variety of food their bodies need to stay healthy. If they do not, they can become ill.

Rickets

Rickets is an illness. It was once common but now it is rare.

When children do not have enough calcium, vitamin D and sunlight, their leg bones do not grow properly. The bones cannot carry the child's weight.

Rickets can be cured by giving children dairy foods and fish and making sure they get some sunshine.

Think about...
Why do we need to drink water?

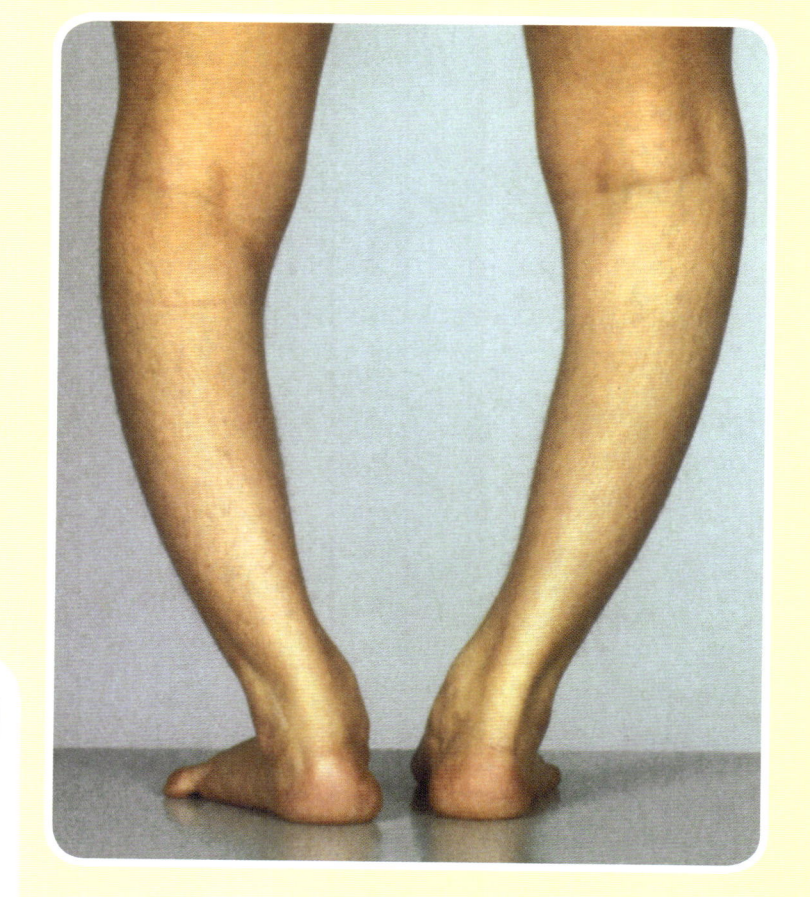

Scurvy

Sailors are at sea for many months. If they do not have a good **diet** they may die of a disease called scurvy. A doctor in the 18th century discovered that eating vitamin C stopped the sailors from becoming ill. Vitamin C is found in vegetables and fruit.

✏️ A woman is worried her child might get rickets. Write a shopping list of foods she should buy to avoid this. Look at the information on pages 120 and 121 to help you.

SHOPPING LIST

✏️ A ship's cook is going shopping to get food for the crew. Write the foods he should buy to make sure the sailors do not get scurvy.

SHOPPING LIST

Diet and exercise

Understand how a varied diet and exercise keep us healthy.

The Big Idea

Animals and humans need water.

✏️ What does this plant need?

Most **living** things are made up of millions and millions of tiny cells. Each cell is filled with a watery liquid. Without water the cells dry up and the life processes cannot take place. Like the plant, we need water so our body can work properly.

Humans need to drink water to replace the water we lose when we sweat, breathe and go to the toilet.

✋ Investigation: How much water do you drink in a day?

Every time you have a drink measure the same amount of water as you drank. Collect all the water together in one bottle or measuring jug.

For example, if you have a glass of orange juice for breakfast, pour a glass of water into the jug.

✏️ How much water have you collected at the end of the day?

💬 Compare your amount with the amounts others in your group collected. Do you drink about the same amount?

Amazing fact
About 70% of our body is made up of water!

124

Amazing fact

780 million people in the world do not have clean water.

It is not always easy for people to get fresh drinking water. Some people have to walk or travel a long way to find water for their families. Some people have to drink dirty water, which can make them ill.

Work out how much drinking water a family of five needs. Use the information from your investigation to work it out.

For some people this glass of dirty water might be the only water they have. Discuss ways to make the water cleaner.

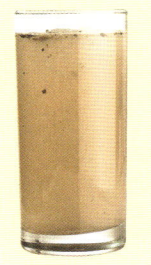

Digging wells is one way of getting clean water to drink.

You can use filters and filter paper. This is a good way to remove the dirt, stones and plants from the water. However, there are millions of microbes (tiny cells) still in the water which are so tiny you cannot see them. These cells can be dangerous. Water has to be treated with chemicals to remove all the microbes. Then the water is safe to drink. Never drink water that you are not sure about even if it looks clean.

Think about...
Are some animals better at storing water than humans?

Amazing fact

Experts agree that about 2–3 litres of liquid a day is enough for most adults.

Diet and exercise

Understand how a varied diet and exercise keep us healthy.

The Big Idea

We can choose foods to make a healthy meal.

Can you remember what type of food gives us energy?

This athlete needs to have a special diet to make sure she is very fit when she competes in competitions. You are going to design a diet for her, thinking about the kinds of foods she needs most.

To help you, first answer these questions. Look at the food plate on page 120 if you need to.

1 What foods will give the athlete energy?

2 What foods will make sure she stays well?

3 What foods will make sure her bones are strong?

Think about...

Why do we need to **exercise**?

Amazing fact

An Olympic rower needs to eat 6 000 kilocalories of food a day when training. That is three times more than the kilocalories recommended for a 9-year-old!

Choose foods and drinks from the word bank to make up menus for the athlete's breakfast, lunch and dinner. One has been done for you. Cross out the words you write on the plates and circle the words you think are not good for an athlete's diet.

Breakfast: yogurt

Lunch:

Dinner:

Word Bank

water   eggs   ice-cream   chocolate   pasta   rice   peas
bananas   carrots   oranges   grapes   couscous   falafel
fish   kebab   milk   lamb   fizzy drink   beans   cereal
khunafeh   burger   khubz   ~~yogurt~~   salad   bread

Write a list of the foods you circled.

Compare your list of foods with others in your group. Are they the same or different? What kinds of food are they?

Compare your plates with your group. Are they similar or different?

Keeping Healthy

127

Diet and exercise

Understand how a varied diet and exercise keep us healthy.

The Big Idea

Exercise keeps us fit and healthy.

Some types of exercise are running, walking, swimming and karate.

💬 What exercise do you like doing?

Many people, especially children, have active lives. This helps them to keep healthy. As people grow older they often become less active. Some adults have jobs where they sit in the same position for a long time.

Movement is one of the life processes so we must move to live. We get energy from our food. We use energy to live, work and exercise. We need to balance the energy we take in with the energy we use. This keeps us healthy.

✋ Investigation: What happens to our bodies when we exercise?

1 Before you begin, sit quietly and relax.

2 Exercise for two or three minutes. Walk quickly or jog, run on the spot or do star jumps.

3 As soon as you stop, sit down.

4 How do you feel compared to when you were sitting down before the exercise? Discuss this with the group.

5 Observe the effects on another student.

💬 Can you see what is happening to the student's chest?

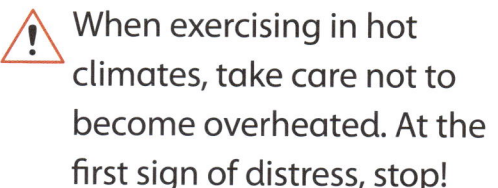

⚠️ When exercising in hot climates, take care not to become overheated. At the first sign of distress, stop!

✏️ What did you find out? Complete the sentences.

I am breathing _____

My heart is _____

I am feeling _____

Amazing fact

An adult's heart beats 60 to 70 times a minute. This can rise to 150 beats a minute during exercise.

Our heart has muscles which pump blood to every part of our body. The blood contains energy from our food and oxygen from our lungs. When we exercise our body needs more energy and oxygen. That is why our heart beats faster and more strongly.

Our lungs absorb the oxygen we breathe in. When we exercise our lungs need to breathe in as much oxygen as possible. That is why we breathe more quickly during and after exercise. The muscles in our chest are working hard to help the lungs breathe in more air.

We need strong muscles so that our heart and lungs can work every minute of every day. Exercise helps us have strong muscles.

How fast the heart beats is called the heart rate. The time it takes for our heart rate to return to normal after exercise is called the recovery rate. We can measure the recovery rate by timing our heart rate. We know we are getting fitter when our recovery rate gets shorter.

✏️ 1 Which group of foods give you energy?

2 Name a food that is a carbohydrate.

3 Name two ways that you can keep healthy and live longer.

4 What does exercise do to the muscles of your heart?

5 What can we measure to find out how fit we are?

Now turn to pages 134–135 to review and reflect on what you have learned.

Damaging foods

Learn that some food can harm our bodies.

The Big Idea

We need to eat foods that are good for us and not just the foods we like.

We need to eat some fatty foods to keep us warm and to give us energy, but we cannot eat too many because this will harm our bodies.

We learned that our heart is very important. It pumps blood around our bodies. If we eat too many fatty foods, the tubes (arteries and veins) in the heart become blocked with fat so the blood cannot flow. When the blood stops flowing we have a heart attack.

Eating the right foods and exercising can protect our heart so it continues to pump well and keep us healthy.

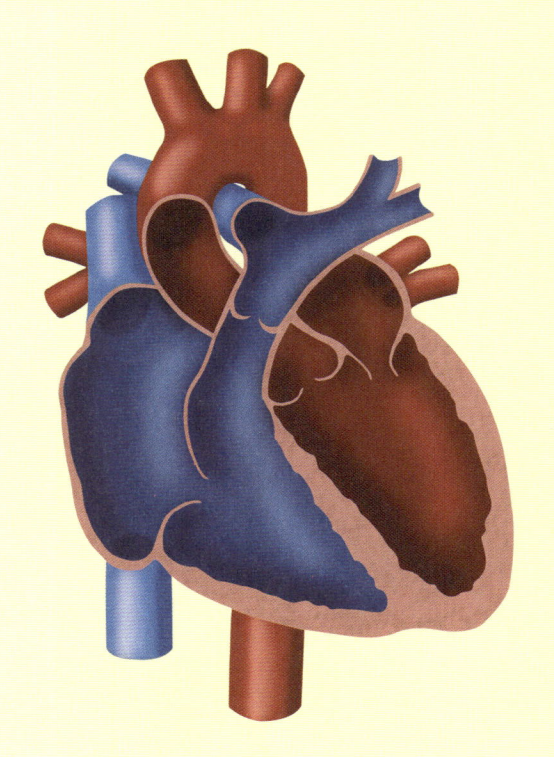

Look at the lunch box. Which of the foods do you think are healthy and which are unhealthy? Write them in the table.

Healthy food/drink	Unhealthy food/drink

✏️ Design a leaflet for a hospital waiting room in your Investigation notebook. Explain why adults should exercise and eat healthily to avoid a heart attack. You can draw pictures too. Think about all the things you have learned about exercise and healthy eating. Use the words in the word bank to help you.

Word Bank

fruit   vegetables   fat   exercise   heart   eat   important

Another common illness is type 2 diabetes. A part of our body called the pancreas produces insulin. Insulin helps our body to use sugar from our food to make energy. If someone eats too much and does not exercise, it can be difficult for the pancreas to produce enough insulin. The amount of sugar in the blood increases and the person can become ill. Exercising and eating healthily helps us to avoid type 2 diabetes.

Type 2 diabetes and heart attacks usually happen when a person gets older but it is important to always eat healthily to avoid getting ill.

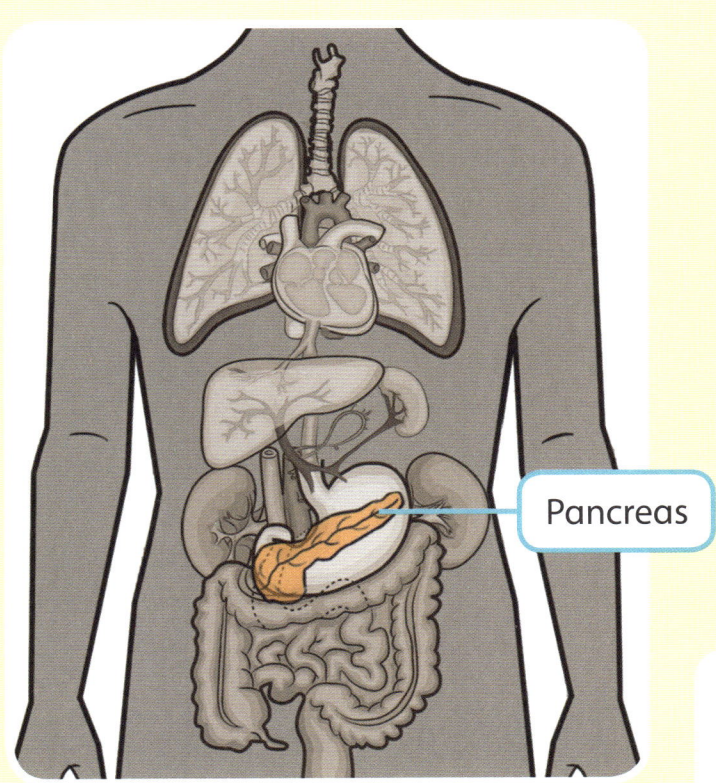

Pancreas

Amazing fact

Worldwide 17 million people died of heart disease in 2008.

Think about...
Are all drinks good for us?

Keeping Healthy

131

Damaging foods

Learn that some food can harm our bodies.

The Big Idea

We should look after our teeth.

💬 What does a dentist do? Do you go to the dentist?

A dentist looks after our teeth. A dentist usually tells us to not eat too many sweet foods and drinks because they are bad for our teeth.

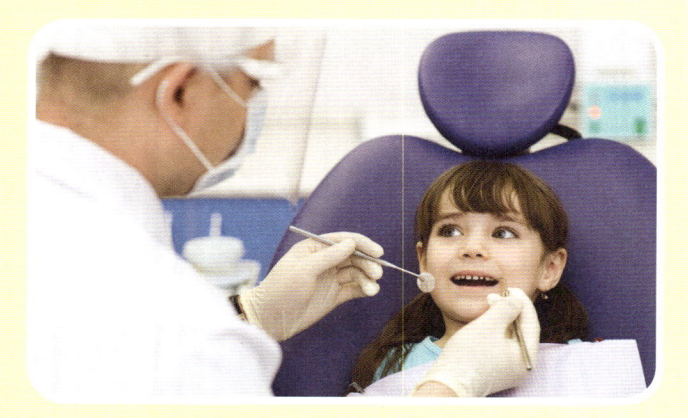

✋ Investigation: Do sweet drinks damage our teeth?

1 Place small amounts of marble or limestone in different drinks. Use water, apple juice, lemonade and cola.

2 Leave the rock pieces in the containers for about two weeks. Take out each piece, dry it and look at it carefully.

3 Record your observations in your Investigation notebook. Draw a picture of each rock at the start of the investigation and after two weeks in the liquid.

✏️ Label the containers with the names of the liquids you used.

💬 Why was it important to place one piece of rock in water?

Why did we use pieces of rock?

✏️ Which drinks caused the most damage to the rocks?

When we drink sugary drinks, microorganisms in our mouth can change the sugar into acids. Some drinks such as cola and lemonade are acidic and start to attack our teeth immediately.

132

How can someone with bad teeth look after his or her teeth better?

Write an email to the person giving some advice. Use the pictures to help you. Copy and complete the email in your Investigation notebook.

Amazing fact

We have 20 baby teeth. We start to grow our adult teeth at about six years old and we lose our baby teeth. We have 32 adult teeth.

Email to the person with tooth decay

From _____

We need to look after our teeth because when we have our adult teeth, we do not grow any more teeth. We use our teeth to chew our food so we can digest it more easily. Without teeth we would have to eat liquid food.

1 Look at page 117 where you wrote down what you like to eat for breakfast. After reading about healthy and unhealthy foods, will you change what you eat for breakfast?

Write your new breakfast menu in your Investigation notebook.

'I will eat… I will drink…'

Explain why you will or will not change your breakfast.

2 Unjumble the letters to find the illnesses we might get if we do not eat healthily and exercise.

a e a h r t   t t a k c a

b d a b i t e s e

c o o t h t   c a d e y

Now turn to page 135 to review and reflect on what you have learned.

What we have learned about keeping healthy

The life processes

✏️ What is nutrition?

✏️ Two life processes are eating and drinking. Can you name two more?

I know the life processes. ○

I know that some people in the world do not have enough food and water. ○

Diet and exercise

✏️ Dieticians put foods into groups: healthy foods and foods that are not so healty.

a Name three foods that are good for you and you can eat lots of.

b Name three foods that are not good for you.

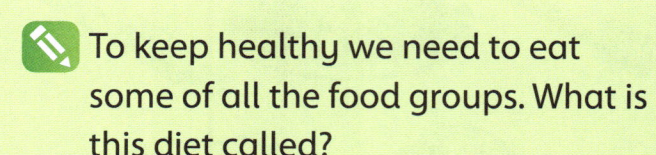

✏️ To keep healthy we need to eat some of all the food groups. What is this diet called?

I understand that some foods are healthy and other foods can be unhealthy. ○

I know that to keep healthy we need a balanced diet. ○

I know how to find out if a meal is part of a balanced diet. ○

134

✏️ Name one way that exercise helps to keep us healthy.

✏️ What do our lungs do?

✏️ How can you find out how fit and healthy you are?

You can measure your _____ rate.

I know that exercise helps us to keep healthy and live longer. ○

I understand why exercise helps to keep us healthy. ○

Damaging foods

✏️ Why might someone get a heart attack?

✏️ What does the pancreas produce?

✏️ What decays the teeth? Circle the correct word.

acid   **sugar**

✏️ How many teeth does an adult have?

I know that some foods are not good for us. ○

I know how to look after my teeth. ○

Keeping Healthy

135

Glossary

Key words

absorbent

animal

diet

exercise

feed

136

feel

flower

food

force

friction

group

grow/growth

Glossary

137

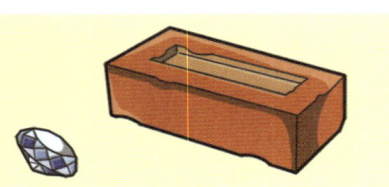

hard

healthy

hear

human

leaf

living

magnetic

material

move

name

nutrition

object

pull

push

Glossary

139

reproduce

root

see/sight

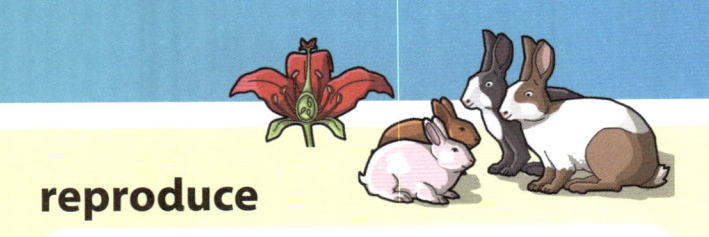

senses

smell

soft

sort

start

140

stem

touch

stop

water

taste

weight

141